Your Older Cat

Your Older Cat

A Complete Guide to Nutrition,
Natural Health Remedies,
and Veterinary Care

SUSAN EASTERLY

APPLE

Published in the UK in 2002 by

Apple Press
Sheridan House
112-116A Western Road
Hove
East Sussex BN3 1DD
UK

ISBN 1 84092 356 3

Designed by Jill Feron
Illustration by Todd Bonita
TMBONITA@aol.com

Printed in China

ACKNOWLEDGMENTS

My deep appreciation and thanks go to all the veterinarians and animal behaviorists who generously shared their time and expertise, particularly James R. Richards, D.V.M.; Shawn Messonnier, D.V.M.; Jill A. Richardson, D.V.M.; and J. Michael McFarland, D.V.M. Without your assistance, this book would not have been written. Any technical errors are my own.

In addition, I would like to thank writers Mary Lidden and Pat Miller for their contribution to the exercise and play section of this book. Special thanks to author Rita Reynolds, whose love of older animals knows no bounds, and to Wendy Simard, for her valuable editorial direction. To fellow members of the Cat Writers' Association (you know who you are), especially Karen Commings and Nancy Marano, I thank you for your encouragement and help. Finally, loving thanks to my family—human and animal—for their great patience and faithful support. My apologies to anyone I have overlooked.

CONTENTS

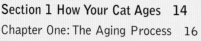

FOREWORD

BY
GERARD K. BEEKMAN, D.V.M.

When I graduated from veterinary school, a "short" quarter-century ago, cats were just a minor species, rarely given more than the last 10 minutes of an hour-long lecture. As pets, they were often relegated to the role of outdoor pest controllers, and they rarely achieved the status they enjoy today. Health care was minimal, nutritional information was lacking, and life expectancies were often limited by disease, accidents, or encounters with other animals, including other cats.

During the last decade, however, cats have attained new popularity as the number-one animal companion in the United States. Longer working days, single-parent households, and the urbanization of the workforce have contributed to the popularity of a creature that can live within a home, often never venturing beyond its walls. The simplicity of dealing with a cat's daily needs, along with its egalitarian spirit and entertaining personality, makes more and more Americans choose felines as their new companions.

Along with changes in the owner's lifestyle, cats have enjoyed changes of their own. No longer second-class citizens, they now benefit from advances in nutrition and disease prevention. Today, veterinarians treat their patients with the concept of wellness in mind. Wellness simply means providing care that optimizes cats' lifestyles so they can maintain consistent good health and a long life. With this kind of care, owners can feel confident that their cat will enjoy a high quality of life well into their golden years.

We all have fallen in love with a cuddly kitten who soon becomes a rambunctious adolescent. That same spry young cat will enter middle age calmer and more reserved. And as your cat begins to age, he may begin to face the first real health challenges of his adult life. Just like aging humans, aging cats can experience more health problems than their younger relatives. The problems are familiar—arthritis, vision and hearing loss, even cancer. But the growing senior cat population (which nearly doubled between 1983 and 1996) has changed the face of feline veterinary care. Though it may be difficult for owners to adjust to this new stage in their cat's life, "getting old" doesn't mean that a cat is dying. Advances in nutrition, medical and behavioral care all contribute to the longevity many cats (and owners) now enjoy. With conscientious care and an eye for prevention, owners can help their cats enjoy long, healthy lives.

LIVING LONGER,
LIVING BETTER

"How old is your cat?" *Cat Fancy* magazine asked in its search for the oldest cat in America. In the months preceding the June 2001 issue announcing the winner, 316 entries for senior cats turned up in the magazine's editorial in box. Some surprising numbers soon emerged: More than half of these cats were 18 or more years old, and a third were 20 or older. The oldest was Starbuck, a handsome 27-year-old striped male cat with bright blue eyes and a hearty appetite. The California resident's veterinarian pronounced him to be "in very good shape," and his lifelong guardian attributed a large part of Starbuck's longevity to "good care and genetics."

Today it is not unusual for a healthy cat to live well into her teens and beyond. Veterinarians across the country report seeing an increasing number of senior cats in their practices, a trend that is sparking new interest in feline geriatric care. Thanks to progress in preventive medicine, nutrition, and knowledge of animal behavior, older cats are blowing out more candles on their birthday cakes. Despite these wonderful advances, the most important part of your senior cat's life, as Starbuck's best friend proves, is you. A well-cared-for older cat can give you just as much pleasure as he did when he was a kitten—as long as he maintains his vitality. Whether you've lived with your aging cat since he was 8 weeks—or 8 years—old, it's up to you to make the most the most of your cat's later years.

CHERISHED RELATIONSHIPS

People began forging relationships with cats more than 4,000 years ago. Over the centuries, cats have survived both our persecution and our adoration. In modern times, the close bond between us is being explored, acknowledged, and respected. Nowhere is this more apparent than in the cherished relationships between senior cats and their human companions.

Today more people live with older cats than ever before. The Pet Food Institute (PFI), which represents companies producing 95 percent of the commercial cat and dog food made in the United States, began keeping track of the number of pets in 1981. Since that time, the feline pet population has jumped by more than 30 million cats! PFI confirms that the number of pet cats reached an all-time high in the year 2000—more than 75 million cats, with over 34 percent of all households living with at least one pet cat. In addition, an American Veterinary Medical Association (AVMA) study shows the percentage of pet cats 6 years of age and older increased from 24 percent in 1983 to just over 47 percent in 1996. There's no denying our graying pet population now lives longer.

Another recent pet study (conducted by the Healthy Pets 21 Consortium, an alliance including the American Animal Hospital Association, the Delta Society, Ralston Purina, and the Center to Study Human-Animal Relationships and Environments, among others) called "The State of the American Pet," reports that most cat owners hug or pet their cat daily (91 percent of 1,000 cat owners). The study also noted that Americans list companionship as their number-one reason for having pets. Most respondents (97 percent) said

In return, evidence supports the idea that cat companionship provides meaningful comfort to us, offering protection against depression and loneliness. At least one study indicates that living with a pet may lengthen human life. Of course, those of us who live with older cats don't need surveys and studies to tell us how much we love our cats or how much they add to our lives. We already know how lucky we are to share this precious time with our feline companions.

"Older cats hold wonderful surprises and gifts for the people they live with and love," says Nancy Marano, a writer who lives with two mature cats in Albuquerque, New Mexico. Marano recently conducted an informal inquiry of people with older cats. Here's what she discovered:

"The bond between the person and cat deepens dramatically. You have accommodated each other for so many years that living together becomes very comfortable. You know each other's moods and needs without guesswork or false steps. You just look at your cat and know if something is wrong or amiss, even if he tries to hide it.

their pet's overall health was good, and on average, cat owners take their cats to the veterinarian 1.6 times a year.

A LITTLE TLC GOES A LONG WAY

Not surprisingly, people who closely interact with their older cats tend to be aware of their pet's needs—physically, mentally, and emotionally. These pet guardians are more likely to seek veterinary care, purchase nutritious cat food, and groom their cats. They take time to play with their cats and keep them safe.

"By the time a cat reaches a certain age, he has you wrapped around his paw. He's trained you to do all the right things. You know where he wants to be rubbed and scratched, how pillows should be arranged on the bed, where his favorite sunning spots are, and when he doesn't want to be disturbed. Your cat doesn't have bursts of kitten energy—but he doesn't knock over lamps, either. It's like a good, long marriage where words aren't necessary to know the depths of the other person."

Just as the aging baby-boomer population causes changes in health-care systems and social-security programs, an increasingly older pet cat population deserves notice as well. This means that as your cat moves into middle age and beyond, you may need to change the way you treat him. If you've picked up this book, you may want to learn more about the joys and challenges of living with an older cat. Perhaps you are interested in discovering simple ways to make your senior cat more comfortable and increase her quality of life. You may be concerned about dealing with a chronically ill cat, considering holistic healing therapies, or seeking solutions for common health problems. You may simply want to understand the feline aging process, from nose to tail. You'll find it in *Your Older Cat*. Our cats are living longer. With this book, you can help them live better.

"Older cats hold wonderful surprises and gifts for the people they live with and love."

10 Cat Wishes for Responsible Pet Owners

1. My life is likely to last 10 to 15 years. Any separation from you will be very painful.
2. Give me time to understand what you want of me.
3. Allow me to place my trust in you— it is crucial for my well-being.
4. Don't be angry with me for long, and don't lock me up as punishment. You have work, your friends, and your entertainment. I have only you.
5. Talk to me. Even if I don't understand your words, I understand your voice when it's speaking to me.
6. Be aware that however you treat me, I'll never forget it.
7. Don't hit me. Remember that I have teeth but I choose not to bite you.
8. Before you scold me for being lazy or uncooperative, ask yourself if something might be bothering me. Perhaps I'm not getting the right food, I've been in the sun too long, or my heart is getting old and weak.
9. Take care of me when I get old. You, too, will grow old.
10. Stay with me on difficult journeys. Never say, "I can't bear to watch it," or "Let it happen in my absence." Everything is easier for me if you are there.

—Author unknown

SECTION 1

HOW YOUR CAT AGES

On a Cat Aging

He blinks upon the hearth-rug
And yawns in deep content,
Accepting all the comforts
That Providence has sent.

Louder he purrs and louder,
In one glad hymn of praise
For all the night's adventures,
For quiet, restful days.

Life will go on forever,
With all that cat can wish;
Warmth, and the glad procession
Of fish and milk and fish.

Only—the thought disturbs him—
He's noticed once or twice,
That times are somehow breeding
A nimbler race of mice.

—Sir Alexander Gray, Scottish professor and poet,
from Gossip, 1928

CHAPTER ONE

The Aging Process

Cats age better than we do. They don't worry about wrinkles or fret over thinning hair. Perhaps in their quiet wisdom they realize aging is a natural process and not a disease. Cats are individuals and, just like people, age in their own ways. Setting a date for old age can be arbitrary and even a bit controversial. In general, veterinarians consider a cat to be a senior beginning around age 7. This does not mean your 7-year-old cat is old; rather, he's at a midlife benchmark. Seven is the age when metabolic changes begin to surface in mature cats, with noticeable physical changes occurring somewhere between 7 and 10 years. Most cats are considered elderly at age 12, but, again, the label can be misleading. The average feline life lasts about 15 to 18 years, but cats who live well into their 20s are no longer considered rare or unusual.

Lifestyle, environment, and whether a cat is spayed or neutered, along with the general wear and tear of daily life, all play a part in the aging process. We may know people who seem younger than their age, and others who look older than their years. Similarly, a stray cat who has endured a hard life may appear older than a pampered housecat who ages gracefully. The same factors that contribute to human longevity—including a well-balanced diet, moderate exercise, and weight control—also can extend a cat's life. A cat who eats nutritious meals, maintains a normal weight, plays, and lives in a healthy environment can enjoy an active life long after she is considered "old." This is especially true when human companions recognize and respond to their cat's changing health needs.

WHAT YOU CAN DO

The aging process occurs gradually, so it's easy to overlook signs of change. Now, before problems begin, is a good time to pay attention to your cat's normal patterns. Gauge his sleeping, eating, and playing habits. Make mental notes of your cat's physical appearance and behavior. Better yet, write your observations down and store them in your cat's health file. Someday your veterinarian will thank you for the feline history notes. In addition, you can do a few simple things on a daily or weekly basis to keep tabs on your senior cat's general health. The Cornell Feline Health Center and the American Association of Feline Practitioners (AAFP) recommend performing a mini health exam as an extension of the way you normally interact with your cat.

> ### Did You Know?
>
> Small dogs live longer lives than large dogs, but this isn't the case with cats, whose body shape remains roughly the same from breed to breed. While a cat's breed does not affect the rate at which he ages, genetic predispositions to different diseases in some cat breeds can result.

- While stroking your cat, check his skin for odd lumps, bumps, sores, and dryness.

- Brush your older cat's fur to remove loose hair, assist grooming, and to stimulate blood circulation and sebaceous gland secretions. Is his fur shiny and soft, or dry and lackluster?

- If she cooperates (try scratching her chin), check your cat's teeth and gums by gently raising her upper lips with your thumb or forefinger. Gums should be pink. Red gums, tartar (a thick coating on your cat's teeth), bad breath, or a broken tooth warrant a vet's attention.

- Look in his ears and examine his ear canals. Healthy ears are clean and odorless.

- Monitor your cat's weight. Run your hands gently over her ribs. Do they feel bony? Try weighing her monthly on a scale sensitive enough to detect small changes. (You can use a kitchen baking scale with a large capacity bowl for very small or slender cats of 7 pounds or less; or weigh yourself on your scale and then hold your cat, weigh again, and subtract the difference.) If there's a significant variation—a pound or more—contact your veterinarian.

• Face him. Are his eyes clear and bright? Any discharge from the eyes or nose indicates a health problem.

AGE OR ILLNESS?

Just like older people, cats tend to sleep longer, eat and drink less, and tire more easily as the years pass. They may experience changes in sight, smell, and taste, and be more sensitive to stress. Vision and hearing may deteriorate. Never assume, though, that changes you notice in your senior cat are simply due to old age and can be ignored. While aging is a normal process, illness is not. Only your veterinarian can tell the difference, and sometimes even she may have a tough time distinguishing between the two. If you notice any physical or behavioral changes, alert your veterinarian.

Unfortunately, cats hide disease well. This normal protective mechanism is an inheritance from their wild ancestors. Cats will hide signs of illness that make them appear weak, even from human companions they know and trust. Cats with chronic disease can look healthy until the illness becomes very serious. Luckily, many common feline diseases are treatable if caught early enough. Do not dismiss significant changes in appetite, weight, elimination, or water intake, even if your cat seems fine. Other red flags are changes in behavior, hearing, walking, or breathing. Early detection especially benefits older cats, whose age-related susceptibility to bodily changes can make them vulnerable to disease within a short time.

PREVENTION IS THE BEST MEDICINE

The AAFP and the Academy of Feline Medicine (AFM) have developed senior care guidelines to promote optimal health in older cats by setting minimal standards of care for cats with and without clinical signs of disease. The guidelines recommend initiating a preventive health-care program for cats between 7 and 11 years of age, one that should continue for the duration of the cat's life. Older cats also should receive a thorough physical exam every six months, a recent departure from traditional care standards.

During the half-yearly checkup, you can expect your veterinarian to assess your cat's general

The Aging Process:
Human Years and Cat Years

You can generally correlate your cat's age in human years using the chart below. A 1-year-old cat is physiologically similar to a 16-year-old person; age-wise, a 2-year-old cat resembles a 21-year-old human. After the age of 2, each cat year equals about 4 human years.

Cat 1 • Human 16

Cat 10 • Human 53

COMPARATIVE AGES
OF CATS AND HUMANS

Cats	Humans
1	16
2	21
5	33
10	53
11	57
12	61
13	65
14	69
15	73
16	77
17	81
18	85
19	89
20	93

Cat 13 • Human 65

Cat 16 • Human 77

Cat 19 • Human 89

physical condition and weight. A complete medical and behavioral history will be gathered, along with a thorough physical evaluating every organ system. The results will then be compared to previous evaluations. At least once a year certain tests—including blood tests, urinalysis, and fecal exam—will be suggested.

Prevention—and early detection—is the key. Catching and treating disorders early, as well as monitoring ongoing medical conditions, is the aim of veterinarians who want to keep your older cat healthy.

Do not dismiss significant changes
in appetite, weight, elimination,
or water intake, even if your cat seems fine.

Healthy-Cat Checklist

If you can't answer yes to all of the following statements, call your veterinarian as soon as possible.

My cat:
- acts normally; seems active and in good spirits
- does not tire easily with moderate exercise
- does not have seizures or fainting episodes
- has a normal appetite
- has had no significant change in weight
- has a normal level of thirst and drinks the usual amount of water
- does not vomit often
- does not regurgitate undigested food
- has no difficulty eating or swallowing
- has normal-looking bowel movements
- defecates without difficulty
- urinates in normal amounts and with normal frequency; urine color is normal
- urinates without difficulty
- always uses a clean litter box
- has not developed new offensive behavioral tendencies, such as aggression or urine-spraying
- has pink gums with no redness, swelling, or bleeding
- does not sneeze and has no nasal discharge
- has bright eyes that are clear and free of discharge
- has a coat that is full, glossy, and free of bald spots and mats; no excessive shedding is evident
- doesn't scratch, lick, or chew excessively
- has skin that is not greasy and has no offensive body odor
- is free of fleas, ticks, lice, and mites
- has no persistent abnormal swelling
- has no sores that do not heal
- has no bleeding or discharge from any body opening
- has ears that are clean and odor-free
- doesn't shake his head or scratch his ears
- hears normally and reacts as usual to his environment
- walks without stiffness, pain, or difficulty
- has feet that appear healthy, with claws of normal length
- breathes normally without straining or coughing

—Reprinted with permission from the Cornell Feline Health Center,
Cornell University's College of Veterinary Medicine

CHAPTER TWO

Season of Changes

The cooler, shorter days, colorful leaves, and dusting of snow during fall and winter signal the aging of the year. These seasons are no less important, or less beautiful, than spring and summer; they are simply part of the year's life cycle. As your cat ages, she also will show physical signs of change; some obvious, some subtle. She will devote more time to sleeping in your lap and less time to chasing balls across the floor. Her coat may thin and the iris, or colored portion of her eye, may seem less vibrant. Eating habits may shift and water intake decrease. If, in her younger years, she couldn't watch a spider scuttle by without batting a paw, she may now be content to watch it spin a web.

BODY TOUR: HOW YOUR OLDER CAT AGES

Unfortunately, your older cat can't describe the age-related changes she experiences. A good way to understand your cat's season of changes is to take a quick tour of her outer and inner workings. Here is what happens as your cat's body ages.

Eyes

The eyes may display the most visible age-related changes. Stare into your older cat's eyes and you may notice a lacy, or moth-eaten, appearance in the iris where the formerly solid, dark swatch of green, blue or gold once appeared. Cats with nuclear sclerosis may develop a bluish cast in the eyes. Iris atrophy and nuclear sclerosis are common age-related traits that may not decrease your cat's vision to any appreciable extent, though several diseases—especially those related to high blood pressure—can seriously impair a cat's ability to see.

Ears

Hearing loss commonly occurs in elderly cats for a variety of reasons, including chronic inflammation due to infection. As a result of hearing loss, your cat may vocalize more loudly and may not respond when you call him. Sudden moves, noises, or touches may startle him.

Teeth

Paying attention to your older cat's dental health in his younger years will pay huge dividends down the road. Dental disease, which can hamper or prevent eating and cause pain, is extremely common in senior cats. In older cats, some teeth may be missing, or look worn and yellow.

To your aging cat, climbing stairs may
now seem like climbing Mount Everest.

Nose

In healthy senior cats, it's possible a reduced sense of smell may partially cause loss of appetite, though a refusal to eat is more likely associated with dental problems or other diseases.

Skin and Fur

A decline in the metabolic rate can cause your cat's skin to become dry and less elastic. An aging cat's skin thins and blood flow decreases, leaving the skin more prone to infection. The skin may appear flaky, and the fur may become thinner and a little dull or rough. As often occurs in aging dogs, your cat's muzzle may turn gray. Older cats groom less efficiently than younger cats, which sometimes results in hair matting, skin odor, and dermatitis.

Muscles, Bones, and Joints

Older cats lose muscle tone and can appear unsteady or wobbly on their feet. The loss of muscle mass may result in a thinner appearance—or the opposite may happen, as your cat's metabolism slows and your pet puts on weight. Older cats often display stiffness in their joints as tissues lose moisture and cartilage gradually deteriorates. Degenerative joint disease, or arthritis, is common. Bones become brittle. Senior cats may have trouble getting up, lying down, or stepping into their litter box. To your aging cat, climbing stairs may now seem like climbing Mount Everest and his favorite perch on your couch even harder to reach. Aging cats may not enjoy being picked up or cuddled if they experience pain. Finally, grooming may become more challenging as stiffness impedes a cat's ability to bend or stretch.

Nails

The nails, or claws, of an aging cat can become thicker, brittle, and overgrown. A senior cat is less able to retract her claws. Be sure to clip her nails (or have them trimmed) to prevent the nails from sticking to textured materials such as carpet or from growing into her paws.

PHYSIOLOGY OF AGING

Research on the physiology of aging in cats is lacking—most of what is known has been learned by observation of diseases associated with aging. A

Top-10 Cat-Care Priorities

1. Improve oral hygiene.
2. Provide fresh water.
3. Avoid harmful substances.
4. Encourage good manners.
5. Visit the vet regularly.
6. Keep cats safe.
7. Maintain healthy weight.
8. Provide a consistent diet.
9. Brush and trim.
10. Feed according to age.

—Reprinted courtesy of the Iams Company

brief system-by-system overview of age-related changes observed in other species, and believed to occur in cats as well, is listed below, courtesy of the Cornell Feline Health Center and the American Association of Feline Practitioners (AAFP).

Immune System

The aging process is associated with a decline in normal immune function and host defense mechanisms, which means older cats become ill more quickly and experience longer recovery times than younger cats. The weaker immune system of a senior cat is less able to fend off infection and disease.

Central Nervous System

In humans, age-related changes in the brain can contribute to loss of memory and alterations in personality commonly referred to as senility. Similar symptoms are seen in elderly cats; they may wander, meow excessively, appear disoriented, and avoid social interaction.

Cardiovascular System

There is no change in normal heart rate associated with aging. Hypertension, or abnormally high blood pressure, is usually the result of kidney failure or hyperthyroidism, a disease associated with overactivity of the thyroid gland.

Respiratory System

Aging lungs display reduced elasticity, tidal volume (the volume of air passing in and out of body during normal breathing), and expiratory reserve. Expiratory reserve is the volume of air that can be forcefully expelled after a normal, unforced expiration. That is, exhale during normal breathing, then force out as much air as you can; the amount forced out is the expiratory reserve. A diminished cough reflex is associated with aging. Primary pulmonary disease is rarely a cause of mortality in older cats.

Gastrointestinal System

It is not clear whether a significant age-related decline in ability to digest and absorb nutrients occurs in all healthy geriatric cats. Protein synthesis and metabolic functions decline in the aging liver, but most common feline liver problems are not routinely associated with age-related changes in hepatic (liver) function.

Urinary System

The feline kidneys undergo several age-related changes that may ultimately lead to impaired function of the organ. Kidney size and renal blood

Behavioral Red Flags

It's often hard to tell the difference between warning signs of illness or normal signs of aging. Call your veterinarian if your cat displays any of the following signs of illness.

- Coughs frequently

- Loses or gains weight

- Loses appetite

- Drinks abnormal amounts of water

- Frequently urinates

- Fails to use his litter box

- Has trouble walking

- Becomes lethargic

- Increased vomiting or diarrhea

- Experiences constipation

- Has trouble sleeping

- Becomes aggressive or disoriented, or suddenly displays other sudden personality change.

flow decline in cats as part of the normal aging process. Disorders of potassium balance occur frequently in elderly cats. Cats, fortunately, do not experience many of the lower urinary tract changes found in human and canine senior patients.

Endocrine System

Physiological changes in the aging feline thyroid gland are not yet well studied. Impaired glucose tolerance increases as cats age.

FELINE FORTITUDE

It's important to keep in mind that cats are generally strong, healthy creatures. Learning about your cat's age-related changes can help you understand how your cat's body works and what your cat is likely to encounter over the years. Knowledge, combined with preventive care (covered in the following chapters), can help you deal quickly with any health problems that develop. A close relationship with your veterinarian can ensure your older cat remains healthy and happy for many years to come.

PREVENTIVE CARE

*Lead your cat to the
feline fountain of youth—
preventive care can help her enjoy
her senior years and increase
quality of life as she ages.*

CHAPTER THREE

Diet for an Older Cat

Food is at the top of your cat's list of good things in the world. Pop open a can of cat food, pour kibble into a bowl, assemble a homemade feline meal, or reach for a healthy kitty treat, and suddenly your sleepy senior resembles a lively, young Cheshire cat, toothy grin and all. Whether you share life with a take-charge cat, whose "Show me the food!" meow greets you more reliably than your alarm clock, or a quieter being who magically appears at feeding time, it's safe to say your older cat cherishes routine days and regular meals. A creature of habit and timing, she prefers to keep things consistent.

*Physical signs of aging can accelerate
when a cat eats a feline version of human junk food.*

If your cat is active, healthy, and in good shape, you don't automatically need to adjust her diet based on numerical age. While it's essential to discuss your cat's individual needs with your veterinarian, veterinary nutritionists agree that thriving older cats can continue to eat their regular adult maintenance diet well into their senior years. The general rule of thumb is to select a nutritionally balanced and complete diet for your cat's particular life stage and needs. The diet you choose should be formulated according to guidelines established by the Association of American Feed Control Officials (AAFCO), which provides uniform pet-food regulations throughout the country. Feeding test protocols provided by the AAFCO allow pet food manufacturers to furnish proof of nutritional adequacy before marketing their products.

To date, only a few studies address the nutritional needs of aging cats during the last stages of life. Luckily, as the number of senior cats continues to rise, veterinarians are focusing more attention on older pets. Reversing the aging process may be the goal of researchers investigating gene splicing and other technologies, but the most important factor in a good, long life is less dramatic. New research now proves that nutrition supports immune function and serves as an effective treatment for disease—a core belief long held by holistic veterinarians. The veterinary goal is, or should be, to improve the quality of your cat's life, not just keep her living longer.

Nutritional requirements do change during a cat's lifetime, of course, thanks to increases in body fat and decreases in lean body tissues. For example, a senior cat's metabolic rate decreases with the years, resulting in a reduction of caloric needs by 30 to 40 percent in the last one-third of her life. Adjusting your older cat's diet to fit her changing needs can help prevent or delay degenerative diseases associated with aging. Even more, optimal nutrition is now the latest—and one of the most effective—tools to help critically ill companion animals recover their health.

Put simply, good nutrition means more vital years for your cat. Don't underestimate the importance of nutrition in your cat's diet—any old grocery-store brand will not do. Physical signs of aging can accelerate when a cat eats a feline version of human junk food. Not surprisingly, many people notice visible improvement in their cat's health when they begin feeding their cat a diet filled with nourishing ingredients.

To keep your cat looking and feeling his best, feed him the highest-quality cat food you can afford. Typically, the more expensive the food, the better the quality, though this isn't always true. You still need to check labels and be an informed consumer (see sidebar on page 42). Keep in mind that your cat is an obligate carnivore, which means she requires nutrients provided by meat. Cats need to eat more protein than dogs and, unlike dogs, cannot survive without meat-based nutrients. While the feline body can synthesize 12 amino acids found in proteins, 10 essential amino acids must come from a cat's diet. For example, the lack of one amino acid, taurine, can lead to blindness and a heart condition called dilated cardiomyopathy. In addition, your cat's critical requirements for arachidonic acid (an unsaturated fatty acid found in most animal fats) and vitamin A also must come from meat.

But cats do not live by meat alone (and certainly not dog food or table scraps). Your cat's optimal diet consists of a balance of protein, carbohydrates, and fats, as well as vitamins and minerals, all of which address the needs outlined above. Because it isn't easy to create and cook regular, nutritious meals for cats, veterinary nutritionists generally recommend that cats be fed a complete, balanced, high-quality commercial food. The trick is to figure out what constitutes a high-quality commercial food (see sidebar on page 41).

THE MOST IMPORTANT NUTRIENT

Along with nutritious food, always provide your older cat with an abundant supply of clean, fresh water to help his body function properly. An aging cat can dehydrate quickly if he experiences vomiting or diarrhea. Encourage him to drink water by using a shallow bowl, offering cold water, or flavoring the water with an ice cube made of something tasty like tuna broth. Because water is so important to your cat's health, go the extra mile and provide him with filtered or bottled water.

Many people don't realize that cats also obtain water from the food they eat. This is particularly true of canned cat food, which contains large amounts of liquid. Cats who eat canned food drink less water than cats who regularly eat dry kibble. There is some debate concerning the merits of feeding your cat wet (canned) over dry (kibble) food. No formal research proves that one is better than the other for healthy cats, though many cat caretakers recommend feeding older cats canned food for at least a portion of their diet. Your cat or veterinarian may do the choosing for you, or your pet may fare better on one or the other. The most important factor is that your older cat eats the right amount of a nutritionally complete diet.

WEIGHT GAIN AND LOSS

A premium diet created for older cats can help balance your cat's nutrition without adding unneeded calories. (Contrary to some recommendations, protein levels in an older cat's diet should be reduced only if your cat is diagnosed with kidney disease—more on this subject in Chapter 10). Obesity may be the most common nutritional problem veterinarians face with older cats. Feline obesity peaks between 6 and 8 years of age, decreases slightly by 10 years, then declines sharply after that age. If your cat gains weight, a physical exam is needed to rule out health problems. Your veterinarian will then recommend an appropriate senior cat food.

Weight loss is the opposite scenario for many cats in the later senior years. Again, a health check will rule out potential diseases that can cause weight loss or a dental problem affecting your cat's ability to eat. At this point, your veterinarian may advise an "energy-dense" food containing higher amounts of fat with adequate amounts of essential fatty acids. These fatty acids, tantalizing to cats, can also help jump-start flagging appetites.

SPECIAL DIETS FOR AILING CATS

Aging cats with conditions such as diabetes, heart disease, or cystitis—to name just a few—may require short- or long-term special diets available through your veterinarian. Prescription diets are formulated to be highly palatable, though some cats manage to resist them quite well. A special diet will not help a cat who refuses to eat it. Work with your veterinarian to establish the best possible diet for an ill or ailing older cat.

HEALTHY SNACKS

It's okay to offer your older cat, and even your dieting older cat, an occasional treat. Try a healthy, low-calorie snack consisting of fresh fruit (an apple slice, grapes, or small balls of cantaloupe or watermelon) or bite-size pieces of raw vegetables (carrot, broccoli, or green beans). A holistic veterinary secret: Raw foods such as carrots can help keep teeth clean.

You'll probably need to experiment to find a healthy treat your cat is willing to eat, but it's worth the effort. When your cat receives treats, remember to reduce his regular food intake slightly to balance the extra food he eats. Snacks should make up no more than 5 to 10 percent of your cat's regular diet. Limit servings to no more than 1/4 cup. On the next page you'll find a popular recipe for a healthy snack, a favorite among feline taste-testers, from Cat Treats.

SUPPLEMENTS—TO GIVE OR NOT TO GIVE?

We humans love to pop pills, especially if we believe those pills or capsules may help solve a medical problem, reduce the risk of disease, or keep us healthier longer. Not surprisingly, the

> ## Manx Meatballs
>
> 1 lb. extra-lean ground beef or turkey
> 1/4 cup chopped fresh parsley
> 1 egg
> 2 tablespoons ketchup
> 1/2 cup dry breadcrumbs
> 1/4 teaspoon garlic powder
>
> Preheat oven to 450°F. Combine all ingredients in a large bowl and mix well. Shape into small balls, and place on broiler pan rack. Bake until brown, 10 to 15 minutes. Serve at room temperature. Store in the refrigerator.

growing usage of dietary supplements by humans has influenced a similar boom in the pet products industry. The recent flood of diet supplements for pets is the reason the AAFCO created the Novel Ingredients Regulatory Framework Task Force, whose chief job is to research new products and determine appropriate labeling. A selection of popular supplements that pay health dividends for older cats may include the following:

- Antioxidants—Vitamins A, C, E, and the mineral selenium help fortify the immune system, deflect environmental toxins, and reduce the risk of developing some cancers

- Biotin—Aids in digestion, skin repair, muscle formation, cellular growth

- Cleansers and detoxifiers—Greens such as chlorophyll, wheat grass, algae, spinach, and broccoli help strengthen the immune system and cleanse the blood

- Glucosamine sulfate and chondroitin sulfate—Helpful in battling arthritis, these supplements lessen joint swelling, stimulate circulation, and build fluid that lubricates joints

- Omega fatty acids—Omega-3s help itchy skin caused by allergies; Omega-6s can help a dull coat shine.

Also new on the scene are dietary supplements to help your overweight cat lose pounds: Carnitine, now added to cat foods designed for weight loss, can increase fat loss and decrease muscle loss; chomium, a glucose-burner potentially beneficial in feline weight-loss programs if used in small amounts; and the better-known fiber, used in place of carbohydrates and fats.

While supplements can benefit cats, do not add any dietary supplement to your cat's food without consulting a veterinarian, holistic practitioner, or veterinary nutritionist. He or she will help you determine an appropriate dosage based on your cat's age, size, and physical condition. If you are tempted to bypass this advice, be aware that giving your cat vitamins or minerals in incorrect amounts can disrupt the nutritional balance of her diet and may have toxic repercussions. Make sure you help, not hurt, your cat.

Pet-Food Shopping Guide

For your cat's sake, don't let pet food labels intimidate you. The Animal Protection Institute (API), based in Sacramento, California, is a nonprofit animal advocacy organization with 85,000 members nationwide. API recommends that you keep the following guidelines in mind when shopping for commercial cat food.

- Make sure the label displays an AAFCO (American Association of Feed Control Officials) guarantee, preferably one that references "feeding tests" or "feeding protocols" rather than "nutrient profiles."

- Avoid cat food containing "byproduct meal" or "meat and bone meal." These rendered products are the most inexpensive sources of animal protein and are not reliable sources of nutrition.

- Look for a specifically named meat or meal, such as lamb or chicken meal, as the first ingredient.

- Avoid generic or store brands. These may be repackaged rejects from big manufacturers and generally contain cheaper, poorer quality, ingredients.

- Bypass "light," "senior," "special formula," or "hairball formula" foods unless specifically recommended by your veterinarian. These foods may contain acidifying agents, excessive fiber, or inadequate fats that can result in skin, coat, and other health problems.

- In general, select brands promoted to be "natural." Look for cat foods preserved with vitamins C and E instead of chemical preservatives like BHA, BHT, ethoxyquin, and propyl gallate.

- Check the expiration date to ensure freshness.

- When you open a bag of dry food, sniff it. If the food smells rancid (bad or "funny"), return it immediately for an exchange or refund.

- Store dry cat food in a sealed nonporous container in a cool, dry place. Remove canned food from the can and refrigerate in a glass or ceramic container.

—Reprinted with permission of the Animal Protection Institute

Pet-Food Label Rules

The 95 percent rule: If the product label says "Salmon Cat Food," 95 percent of the product must be the named ingredient.

The 25 percent or "dinner" rule: Ingredients named on the label must comprise at least 25 percent of the product (but less than 95 percent) when a qualifying "descriptor" term such as dinner, entrée, formula, platter, or nuggets is used. In "Turkey Dinner for Cats," turkey may or may not be the primary ingredient. If two ingredients are named, they must total 25 percent; there must be more of the first ingredient listed than the second; and the lesser ingredient must be at least 3 percent of the total.

The 3 percent or "with" rule: A product may be labeled "Cat Food with Salmon" if it contains at least 3 percent of the named ingredient.

The "flavor" rule: A food may be labeled "Turkey Flavor Cat Food" even if the food does not contain turkey, as long as there is a "sufficiently detectable" amount of flavor. This may be derived from meals, by-products, or "digests" of various parts from the animal species indicated on the label.

—Reprinted with permission from the Animal Protection Institute

Keep in mind that pet-food manufacturers are often one step ahead of us. For example, antioxidants, known for combating destructive free radicals—cells implicated in a weakened immune system, degenerative arthritis, and a host of diseases from cancer to cardiovascular disease—have hit the mainstream pet-food industry with a wallop. Accordingly, pet-food manufacturers increasingly proclaim the addition of antioxidants to their list of ingredients.

To help ensure you're buying quality nutritional supplements:

1. Look for three important letters on the label: USP (which stands for United States Pharmacopoeia). This not-for-profit organization sets standards for prescription and nonpre-scription medicines, including dietary supplements for veterinary use.

2. Ask your veterinarian or consult a veterinary nutritionist. (Your veterinarian can recommend one or refer you to a holistic veterinarian who specializes in nutritional therapy.)

3. Do some research. Contact a nutritionist at a veterinary college or consult a veterinary nutritionist at the American College of Veterinary Nutrition (ACVN). The ACVN Web site is ACVN.org. If you have a specific question about cat nutrition, you can contact the ACVN office and be referred to a Diplomate who specializes in that area.

CHAPTER FOUR

Exercise and Play
to Keep Your Cat Active

Your cat rouses from a long nap, yawns, and stands on all fours with her feet apart. She exhales while arching her back. Holding this pose for a few timeless seconds, she then inhales as she slowly curves into a concave position. She raises her head and looks up in a luxuriously slow, flowing movement. If you swear your balletic cat just performed a perfectly lovely yoga posture, you're absolutely right. Human yoga postures, called asanas, have been drawn from natural sources for thousands of years. It has been said there are millions of creatures—and as many yoga postures. "The Cat" asana your cat performs strengthens the back and pelvic area, promotes spinal flexibility, and energizes the feline body. It's no wonder that humans adapted this beautiful stretch from cats.

Of course, cats don't need exercise videos to show them how to stay in shape. Keeping fit comes naturally to them, thanks to a never-forgotten need to survive in the wild. And they manage it without doing anything resembling strenuous exercise such as aerobics or long daily runs. What's their secret? You've probably already guessed that stretching is the key. Cats have the wonderful ability to gracefully stretch every part of their body so powerfully that their muscles vibrate. For people and cats, stretching movements can help strengthen muscles, reduce excess weight, and release nervous stress.

A feline preference for warmth (due to its origins in warm climates) is one theory that may help explain why cats stretch frequently. Aging cats,

however, may need a little help to stay sleek and supple, especially if they live in cold climates unkind to stiff joints. You can encourage your older cat to stretch in a warm area, or invite her to join in as you stretch or perform yoga. An enticing scratching post (try a sisal-covered one with a secure base and a small amount of catnip rubbed or scattered on the post) can spark a spine-tingling stretch, or you can simply encourage your cat to reach up to you. An older female cat may enjoy balancing on her front paws while you gently stretch her back paws.

THE ART OF PLAY

Just like us, cats need exercise to lift their spirits and keep them fit. Regular, moderate exercise is the not-so-secret key to weight control and overall

How Does Your Cat Measure Up?

Very Thin

- Ribs, backbone, and pelvic bones easily visible, with no fat cover
- Thin neck and narrow waist
- Obvious abdominal tuck
- No fat in flank folds, folds often absent

Underweight

- Backbone and ribs can be easily felt
- Minimal fat covering
- Minimal waist when viewed from above
- Slightly tucked abdomen

Ideal

- Ribs can be felt, but not visible
- Slight waist line observed behind ribs when viewed from above
- Abdomen tucked, flank folds present

Overweight

- Moderate fat cover over ribs, but can still feel ribs
- Abdomen slight rounded, flanks concave
- Flank folds hang down with moderate amount of fat, jiggles when walks

Obese

- Ribs and backbone not easily felt due to a thick fat covering
- Abdomen distended; waist barely visible or absent
- Prominent flank folds which sway from side to side when walking

Toys to Fit Your Cat's Personality

Wonder what sort of toys match your older cat's personality? Here are some thoughts on the subject, courtesy of CatToys.com. To spark your senior cat's interest in a new toy, sprinkle or rub a small amount of catnip on the toy or in the wrapping paper.

Extremely active cats: If your senior cat springs to life occasionally, try pounceable catnip pillows, an unpredictable rolling activity ball, a bird door hanger, or a realistically squeaky toy mouse.

Moderately active cats: Try soft fabric catnip mice with leather tails; free-standing bird on a spring that rocks back and forth; and motion-sensitive blinking ball with light, bell, and yarn tail.

Couch potato cats: Snacking toys can get your cat off the bed. Consider a hollow ball—the opening at each end allows treats to fall out as he plays. Even couch potatoes love a good feather toy—try furry catnip mice with feather tails.

Plays well with others: Cats who play well with others enjoy interactive play with people too. Choose toys that you and your cat can play with together, or toys that work well for multiple-cat households. Try a feather and felt "chaser" toy packed with catnip, kitty-size catnip pillows to swat or pounce on, or soft fabric catnip mice with leather tails.

Prefers to play alone: Even the most independent aging cat can have a healthy appetite for play. Interest your cat in a reshapeable "messy" toy mouse made of natural sheep's wool with a leather tail, nose, and ears; or try a full-body wrestle (front paws, hind legs) "mollusk" made from fleecy material, topped with feathers and stuffed with catnip.

Insists on playing at night: Nighttime toys for the cat who has a sudden burst of playful energy in the wee early hours include a blinking laser ball; a catnip-refillable, glow-in-the-dark ball; and a flashing toy ball that lights up on impact.

CAUTION: Your cat may be older and wiser, but she still needs supervision during playtime, especially when she plays with toys containing wire, string, or mechanical parts. Remember to put these types of toys safely away when play ends.

health. Increased blood circulation and improved muscle tone are precious gifts to older cats, who become less mobile as arthritis sets in and muscles atrophy. People jog, bike, or swim, but a cat's best activity is playing with you. The act of playing comes naturally to kittens instinctively developing survival skills, but older cats who mastered these techniques long ago (and haven't found much use for them in the meantime) may need more encouragement to get moving. This isn't usually a problem if you've played together since your longtime companion was a kitten. Give him toys that invite him to stalk, climb, leap, and pounce.

These playthings can run the gamut from the simple and cost-free brown-paper shopping bag, cardboard box, or wad of paper on a string to commercially made cat toys. Fortunately, cats are just as content with paper bags and cardboard boxes as they are with expensive or elaborate store-bought toys. If you have a limited budget, tour pet-supply stores for inspiration and make your own.

IT'S NEVER TOO LATE

If you've adopted an older cat or never thought to play with your cat during his early years, it's not too late to start. Begin slowly, perhaps with a strip of paper to encourage the smallest swipe of a paw. Gradually increase the length of time you play with your cat. Schedule regular play sessions at approximately the same time every day to give your routine-loving cat something to look forward to besides meals. Try playing before a meal to kindle her appetite, which simulates feline predatory behavior in the wild, or play in the evening to discourage nighttime activity.

Remember to challenge your cat's brain as well as his body. (This may call for some brain stretching on your part, too.) Positive reinforcement is the key. For instance, choose a simple trick or action you want your cat to perform when asked and consistently reward him every time he does it. The reward doesn't have to be a food treat; reward him with a pat or a verbal "yes!" instead.

Check with your veterinarian before you enlist your cat in a feline exercise program, especially if he is overweight and sedentary or becomes easily winded and tires quickly. Once you get the green light, choose games that are appropriate for your senior's age and condition. Start with simple exercises and slowly build up to more challenging games when your cat appears fit and ready.

10 Steps to Help Your Older Cat Lose Weight

Just like people, older cats often add pounds with the years. You can do a lot to help your plump cat drop those extra pounds and regain her svelte shape. Follow these 10 steps suggested by the Iams Company to help your cat reach a healthier weight. Remember to check with your veterinarian first, though, to establish goals and ensure your cat is otherwise healthy. Follow-up visits can help you and your veterinarian monitor your cat's progress.

1. Let the games begin. Playing can help your cat burn calories. Toss her toys to chase, wiggle a wand for high jumps, or provide a taller cat scratching post to climb.

2. Go for a walk. Take your cat for a walk, even if it's just inside the house. Many cats learn to enjoy walking on a leash, and it's a great opportunity for you to get some exercise too.

3. Ease into shape. Watch how your overweight cat handles increased activity. Don't let him become exhausted, overheated, or out of breath. Older cats may not be able to exercise vigorously.

4. Replace treats with praise. When your cat begs for treats, she may be begging for attention. Substitute play, grooming, stroking, or conversation for food treats as expressions of love. You can also try catnip as a nonfood treat.

5. Resist those pleas. Is your cat an expert at begging for table scraps? Keep your cat in a separate room during your meals if you find it difficult to ignore those pleas or wails.

6. Feed cats individually. If you have more than one cat, consider keeping them in separate rooms during mealtimes. This will prevent the greediest cat from overeating and ensure that the slower-eating cat gets fed.

7. Play fetch. Toss kibble to your cat, one piece at a time, to combine exercise with mealtime.

8. Avoid fiber overload. Many reduced-calorie pet foods include increased levels of fiber that can interfere with a cat's ability to absorb and digest nutrients. Food with the proper balance of animal-based protein, fat, carbohydrates and moderately fermentable fiber sources, such as beet pulp, is a healthier choice.

9. Smaller meals, more often. Feeding your cat several small meals each day, rather than one large serving, can help burn more calories through meal-induced thermogenesis. This process of heat produced by the body during digestion, absorption, metabolism, and storage of nutrients causes more calories to be used.

10. Tip the scales. Keep track of your cat's weight by using a baby scale. Or, take your cat in your arms, step on your scale, and subtract your weight from the total weight shown to find your cat's weight. Check your cat's weight-loss progress every two weeks.

CHAPTER FIVE

Comfort Zone: Making Life Easier

To a great extent, a senior cat's quality of life centers on her ability to continue to get around, reach her favorite places, and live as comfortably as possible. You can contribute to your older cat's physical and mental well-being in numerous ways. As the old saying goes, it's the little things that count.

Visit any well-stocked pet supply store or pet-related Web site, and you may be amazed to find several product lines designed specifically for the needs of older animals. You will spot everything from microwaveable heating pads and orthopedic beds to stylish cotton overalls with disposable pads (for incontinent cats) and specially designed ramps and steps.

Consider the following sample of comfort toys for feline couch potatoes:

Snuggle Kitties. Originally designed for use with foster kittens, this stuffed kitty has a hook-and-loop closed cavity under its belly with a disposable heat pouch that stays warm up to 20 hours, or a cloth pouch that holds rice and can be microwaved to keep your cat warm. Comes with a plastic, battery-operated heart that fits into the under-belly opening and simulates a real heartbeat.

Thermo Kitty Bed. This round bed with tall sides includes a removable dual thermostat heater with a super-soft, orthopedic foam base and removable faux lambskin cover for easy washing. The heater is also removable, plugs into any outlet, and uses only four watts of power.

Cat Sitter Video. Mice, squirrels, birds, and fish entertain your cat in this colorful video available in both one-hour or six-hour versions.

Of course, you don't have to spend a lot of money to make your older cat more comfortable; you just need to look at life from her point of view.

Following are some comforting suggestions for older cats, including those with disabilities:

* For stiff or arthritic cats, create steps or use footstools to help them reach their favorite hangouts or resting places. Securely attach carpet to wooden steps to prevent slipping, and ease her transition over steps or stairways with carpet-covered ramps. A paint tray used as a litter box can provide a painless step into the low side, and an elevated food bowl can relieve pain associated with bending down.

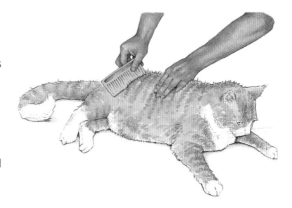

- For cats with cataracts or poor vision, leave a light on during the night to help your nocturnal cat see. Don't routinely pick up a blind cat; he is likely to be disoriented when you set him down.

- For cats with hearing problems, flick a flashlight on and off to help train or "call" your cat. Your hard-of-hearing cat may also respond to a high-pitched dog whistle.

- Piles of pillows can ease a handicapped cat's path to a higher location.

When it comes to making your senior cat more comfortable, you are truly the expert. No one knows your cat, or her needs, like you do. Use the suggestions listed above to launch your own creative ideas.

GOOD GROOMING

Cats enjoy training their human companions in the delicate art of grooming them properly. One lovely 12-year-old calico announces grooming time to her human with a demanding meow and a flop onto her side twice a day. When one side is properly attended to, she rolls over to the other side. She imposes no time limit on this procedure, reluctantly moving along only when the hand brushing her tires. Not all cats enjoy this much pampering, but older cats often need some hands-on help to pick up where their self-grooming duties stop. You'll know your cat needs your help when she begins to look like a person who just got out of bed in the morning—a little sleepy and unkempt. Pay attention to any abrupt change in your cat's grooming habits, though. A sudden lack of groom-

ing may signal the onset of a health problem. Check with your veterinarian if this occurs.

Bathing your older cat isn't necessary unless he becomes dirty. Brushing or a gentle combing every day will make him feel more comfortable in his skin and closer to you. He'll enjoy the attention, and you will do the good work of stimulating his blood circulation and sebaceous gland secretions, removing scales next to the skin and returning a luster to his coat. Brushing also removes dead or loose hairs that otherwise might play a starring role in your cat's next hairball. And hairballs can cause potentially serious problems—like recurring impaction—to aging gastrointestinal tracts.

Finally, daily grooming can help you catch flea or tick problems early. Once-a-month flea and tick "spot" products, available by prescription through your veterinarian, can help prevent any infestation.

DON'T PULL!

Brushing a shorthaired cat usually doesn't present much of a challenge, but the aging longhaired variety can be infamous for mats or clumps of clotted fur. One veterinarian who works with geriatric cats advises working mats out gently and slowly with a wide metal-tooth comb instead of scissors. Take your time, let your cat be your guide, and honor her request to stop. You can always try again tomorrow. Offer her a healthy treat or a nip of catnip and praise her for her patience.

Last but not least, don't forget the bottom line. Unsightly knots can bunch like grapes around your longhaired cat's anal area—a place that many aging cats find tough to clean after a bowel movement. If she's not too insulted by your behavior, try a quick, gentle swab under her tail (baby wipes work well) after she uses the litter box. Remove existing clumps as gingerly as you can, or consult a professional groomer for advice.

TOOTH AND NAIL

Don't forget the other important brushing your senior cat needs. A cat's teeth play a critical role in her life. Dental problems are not only painful, they also are a prime factor in cat behavior problems, illness, and disease. Left untreated, bacteria in a cat's mouth may infect other organs, such as the heart or kidneys, and shorten her life. On the other hand, healthy teeth and gums can significantly extend the quality and length of life.

Veterinarians agree that good dental care is one of the most important things you can provide for your aging cat. Evidence indicates that brushing a cat's teeth and gums daily or weekly reduces tartar formation. Many diligent cat lovers use a piece of gauze to gently rub their cat's teeth and gums

with toothpaste made for cats (don't use human toothpaste on your cat's teeth). If you've never done this and don't want to, or your cat makes it clear her mouth is off-limits, you have other options. Most cats like the feline dental chews available at pet-supply stores, or ask your veterinarian about a prescription food approved by veterinary dentists to help keep teeth clean. Whether or not you brush your cat's teeth, schedule a dental exam. Regular dental checkups—at least once a year—and professional cleanings are vital to maintain your cat's good health.

Finally, check your senior cat's nails regularly. A weekly trim is in order if your aging kitty clicks her nails across the floor. To trim claws, choose nail clippers designed for cats. Grip a paw and press gently on the footpads to extend a claw. Carefully snip the sharp edge just below the quick, or pink area, of the nail. (Cutting into the quick will result in pain and bleeding—apply styptic powder to stop bleeding if this occurs.) If you're not sure where the quick begins on your cat's nails, you can safely clip the tips of her long nails.

SHOULD I SPAY OR NEUTER MY OLDER CAT?

Yes! Although spay-and-neuter advocates generally focus on how young a cat can handle the operation, spaying (removal of ovaries, uterus, and tubes that carry eggs—stitches required) or neutering (removal of both testes and part of the tubes that carry the sperm—no stitches needed) will make your pet more comfortable at any age— and improve your cat's longevity, health, and behavior. Pre-admission blood work, a checkup and other tests deemed necessary by your veterinarian are a must, however, to help ensure your cat is healthy enough to undergo the surgery. Placing a cat under anesthesia can be risky no

matter what his age, but the risk is thought to increase as your cat ages; your veterinarian will take every precaution to minimize negative effects. The benefits outweigh any danger when you consider the alternative: An older female cat can have potentially fatal uterus problems, such as pyometra (pus and bacteria in the uterus); continuing pregnancies (seen in cats up to 15 years old), which take a terrible toll on bones, teeth, and body; and increased susceptibility to breast cancer. Unaltered male cats experience fewer health problems, but they continue to fight, howl, and spray urine to mark territory as long as testosterone rules them, making them less-than-pleasant housemates.

On the other hand, spaying eliminates heat cycles and the often-irritable behavior that accompanies them. Neutered male cats are more affectionate and roam less. Weight gain can be a side effect of spaying or neutering. The operation doesn't cause it, but your cat's newly relaxed attitude may result in a less-active lifestyle. You can avoid this problem by feeding proper amounts of food and making time for daily play sessions with your cat.

REDUCE STRESS

For older cats, a good day is one that is calm and predictable. From your cat's perspective, however, life with humans is often filled with stressful events. The worst offenders include moving to a new home or adding another pet to the household. Here are a few ways to de-stress your cat's environment.

Clean his litter box

Cats believe cleanliness is next to, well, you know. Neat and picky, cats won't settle for anything less

than a clean box. Cat food should not be kept in the vicinity of a litter box. If you must move his litter box to another area, show him the new location — stir the litter to make sure he gets the message.

Keep it quiet

Loud noises or commotion easily scare older cats with diminished hearing or sight. They may be unable to move quickly or get out of the way. Avoid the startle factor and advise others, especially kids, to keep things as calm as possible.

Help her sleep

Aging cats nap more but they may also be subject to sporadic sleep, which can cause you stress. Give your cat a warm, soft bed in a draft-free zone, play classical music before sleep and a provide night light.

Introduce new pets gradually

The jury is out on whether adding a younger cat to the household constitutes positive or negative stress. Some experts believe introducing a new pet is a traumatic experience best avoided; others believe a younger cat can encourage an older one to be more alert and playful. If you elect to introduce a new kitty, be patient, take things slowly, and don't leave the two cats alone together until you are sure they can give peace a chance.

Soothe sudden changes

ometimes a big upheaval, like moving to a new home, is unavoidable. Keep your cat in a quiet area of the new location at first. Try to keep everything else, such as food and bedding, the same as before. Lighten your older cat's stress level during a major transition by giving him extra affection, attention, and playtime with you.

SECTION 3

NATURAL RELIEF

*Preventing disease is a primary
goal of natural health. A gentler, holistic
approach to care can be particularly
advantageous for older cats, because
of the strong focus on nutrition
and supportive therapies.*

CHAPTER SIX

Holistic Healing Therapies

Natural care may be a new concept to many cat owners, but alternative treatments have been around for thousands of years. According to legend, the first nonhuman acupuncture patient was an elephant treated for a stomach disorder about 3,500 years ago. In the twenty-first century, holistic healing therapies for people and animals are gaining ground in the United States for several reasons. The American Holistic Veterinary Medical Association (AHVMA) points out that holistic medicine, by its very nature, is humane to the core. Techniques used in holistic medicine are gentle, minimally invasive, and incorporate patient well-being and stress reduction. Holistic thinking centers on empathy, respect, and loving care. One natural practitioner notes that as more people embrace these therapies for themselves, they naturally want them for their pets.

Holistic veterinarians, who often begin their careers as mainstream practitioners, search beyond conventional Western medicine's focus on fixing or eliminating symptoms of illness to encompass a "bigger picture" of an animal's over-all health and well-being.

Preventing disease is a primary goal of natural health. For older cats, a holistic approach can be particularly advantageous, because of the strong focus on nutrition and supportive therapies.

In fact, nutrition is the foundation upon which holis-tic healing therapies are built. Many holistic veteri-narians and practitioners advocate a well-prepared homemade diet if a pet guardian has the time and commitment to do it properly. A complete under-standing of feline nutrition is essential, however, which means you'll need to do some research first.

Another option is to choose a commercially pre-pared high-quality, all-natural cat food, such as Innova or Wysong, which are available in some pet supply stores or natural food outlets and through mail order.

HOME-PREPARED DIET FOR ADULT CATS

CAUTION: Before you begin feeding your cat a home-prepared diet, discuss the decision with your veterinarian or a holistic veterinarian in your area. The following ingredients, recommended by the Animal Protection Institute (API), make approxi-mately three servings. Feed an adult cat as much as she will eat in 20 to 30 minutes, and refrigerate leftovers promptly. Feed adult cats twice a day.

Choose One Protein Source
(Meat amounts given in raw weight)
1/2 lb. boneless chicken breast or thigh, minced
6 oz. ground turkey or minced turkey (dark meat)
1/2 lb. beef, chicken or turkey heart, ground or minced (About 3 times a week, include 1 chopped hard-boiled or scrambled egg.)

Optional: once a week, substitute 4 oz. organic liver for 1/2 of any meat source

Optional: once every two weeks, substitute 4 oz. tuna (packed in water, no salt), 6 oz. sardines (canned), or 5 oz. salmon (canned, with bones) for any meat source. Do not use canned fish as a pro-tein source for cats prone to urinary tract problems.

Optional: for cats needing a lower protein diet, add 1 cup cooked white rice.

Supplements
- 2 tsp. olive oil, or 1 tsp. olive and 1 tsp. flaxseed oil

- 300 mg calcium (as carbonate or citrate), or about 1 slightly rounded tsp bonemeal (human grade). If using canned fish with bones, decrease calcium to 1/4 regular amount.

- 1 to 2 tbsp. pureed vegetables—many cats pre-fer their veggies lightly steamed—or vegetable baby food (without onion powder!)

- 1/4 tsp. salt substitute (potassium chloride)— give 3 or 4 times a week

*Holistic veterinary medicine
incorporates, but is not limited to,
the principles of acupuncture
and acutherapy, botanical
medicine, chiropractic, homeopathy,
massage therapy, nutraceuticals,
and physical therapy,
as well as conventional medicine,
surgery, and dentistry.*

- 1 cat-size dose of multiple vitamin-mineral supplement (human quality) or cat vitamin

- 1 probiotic/digestive enzyme supplement

- 80 mg taurine (about one 250 mg taurine capsule or tablet, powdered). Omit if using cat vitamin.

IMPORTANT: Go slowly and proceed gradually with any diet change. Your animal companion's tastes and digestive system need time to adjust to new foods. Too rapid a change may result in diarrhea, vomiting, refusal to eat, or other problems. This is especially important for older or sick cats.

HOLISTIC CARE

Holistic veterinary medicine is a comprehensive approach to health care employing alternative and conventional diagnostic and therapeutic modalities, according to American Veterinary Medical Association (AVMA) Guidelines for Alternative and Complementary Medicine. In practice, the guidelines specify that holistic veterinary medicine incorporates, but is not limited to, the principles of acupuncture and acutherapy, botanical medicine, chiropractic, homeopathy, massage therapy, nutraceuticals, and physical therapy, as well as conventional medicine, surgery, and dentistry. The AVMA recommends that holistic veterinary medicine be practiced only by licensed veterinarians educated in the modalities employed.

In 1996, the AVMA revised its guidelines for alternative and complementary veterinary medicine. Below are the organization's brief descriptions of alternative therapies.

Acupuncture and Acutherapy
Veterinary acupuncture and acutherapy involve the examination and stimulation of specific points on

the body of animals by use of acupuncture needles, moxibustion (heat used to stimulate acupuncture points by burning different materials indirectly on the needle or directly on the skin), injections, low-level lasers, magnets, and a variety of other techniques for the diagnosis and treatment of numerous conditions in animals.

Chiropractic

Veterinary chiropractic is the examination, diagnosis, and treatment of animals through manipulation and adjustment of specific joints and cranial sutures. The term veterinary chiropractic should not be interpreted to include dispensing medication, performing surgery, injecting medications, recommending supplements, or replacing traditional veterinary care.

Physical Therapy

Veterinary physical therapy is the use of noninvasive techniques, excluding veterinary chiropractic, for the rehabilitation of injuries in animals. Veterinary physical therapy performed by nonveterinarians should be limited to the use of stretching; massage therapy; stimulation by use of low-level lasers,

electrical sources, magnetic fields, and ultrasound; rehabilitative exercises; hydrotherapy; and applications of heat and cold.

Massage Therapy

Massage therapy is a technique in which the person uses only the hands and body to massage soft tissues. (More on this in the next chapter.)

Homeopathy

Veterinary homeopathy treats nonhuman animals by administering substances that are capable of producing clinical signs in healthy animals similar to those of the animal to be treated. These substances are used therapeutically in minute doses.

Botanical Medicine

Veterinary botanical medicine is the use of plants and plant derivatives as therapeutic agents.

Nutraceuticals

Nutraceutical medicine is the use of micronutrients, macronutrients, and other nutritional supplements as therapeutic agents.

Homeopathic First-Aid Kit

Homeopathy, a complex system of holistic medicine using small doses of natural substances to stimulate the body's natural defenses, operates on the premise that like cures like. In a highly diluted form, homeopathic remedies are said to cure what they might cause in a larger amount, thus stimulating the body to heal itself. Though you should always consult a homeopathic veterinarian for correct dosages before treating your older cat, the following items can be kept on hand for emergencies.

- Arnica for wounds
- Hypericum for traumatic injury/possible nerve damage
- Nux vomica for diarrhea
- Silicea for abscesses due to fights

What to Look For

Searching for a natural practitioner in your area? Look for a licensed practitioner who has received formal training. For example, the International Veterinary Acupuncture Society (IVAS) maintains a list of all certified veterinary acupuncturists. Formed and chartered in 1974, the nonprofit IVAS strives to establish uniformly high standards of veterinary acupuncture through educational programs and accreditation. The following list will help point you in the right direction.

Academy of Veterinary Homeopathy (AVH)
6400 East Independence Blvd.
Charlotte, NC 28212
(866) 652-1590 • www.theavh.org

American Holistic Veterinary Medical Association (AHVMA)
2218 Old Emmorton Road
Bel Air, MD 21015
(410) 569-0795 • www.ahvma.org

American Veterinary Chiropractic Association (AVCA)
P.O. Box 563
Port Byron, IL 61275
(309) 658-2958 • www.animalchiropractic.org

Animal Natural Health Center
1283 Lincoln Street
Eugene, OR 97401
(541) 342-7665 • www.drpitcairn.com

The British Association of Homeopathic Veterinary Surgeons
Chinham House
Stanford in the Vale
Oxon SN7 8NQ
01367 718115 • www.bahvs.com

International Veterinary Acupuncture Society (IVAS)
P.O. Box 271395
Fort Collins, CO 80527
(970) 266-0666 • www.ivas.org

CHAPTER SEVEN

Hands-On Alternative Treatments

Feeding, petting, and holding are all comforts your hands provide. Use this knowledge to help heal your cat: Petting can become a comforting massage, or you and your cat can discover the natural healing power of Reiki, TTouch, aromatherapy, and flower essences. Often used in conjunction with holistic healing methods and conventional treatments, these hands-on modalities help promote peace and healing. In addition, observing your cat's body language can bring understanding and a deeper level of communication between you.

CAT MASSAGE

Treated to gentle handling that considers stiff joints and fragile bones, most aging cats learn to appreciate a good massage. Indeed, this hands-on communication may be one of the nicest things you do for your older pet. Touch is powerful—it conveys love and deepens trust while bestowing physical benefits. Massage can help increase your cat's blood circulation while nourishing tissues and soothing aching muscles.

FLOWER ESSENCES

Used for centuries by primitive cultures, flower remedies were rediscovered in 1928 by Edward Bach, M.D., a British physician, bacteriologist, and immunologist. He is perhaps most well known for his popular Bach Rescue Remedy, which combines the essences impatiens, star-of-Bethlehem, cherry plum, rock rose, and clematis. Rescue Remedy is universally used to reduce stress in people and animals (you'll find it at most health food stores). Holistic veterinarians and cat behaviorists often recommend flower essences for healing feline emotions. Made from essential and highly diluted oils, flower remedies—a few drops at a time—are usually added to water or given orally. They are easy to administer and can be used by themselves or with other healing therapies.

While Bach flowers were originally created for people, Anaflora flower essences are made especially for animals by Sharon Callahan, an author, internationally recognized animal communicator,

and pioneer in treating animals with flower essences. She describes flower essences as non-aromatic vibrational tinctures made from the blossoms of flowering plants. Each plant—the flower is the most potent part—has individual healing properties. To make a flower essence, the healing properties of the plant are transferred to water by a process of sunlight infusion. The resulting essence can be rubbed on the skin, ingested, or sprayed around the animal being treated. Anaflora offers 118 individual flower essences and 32 flower essence formulas that address the physical, mental, emotional, and spiritual needs of animals. (See Resources.)

REIKI

The name for this ancient hands-on healing modality (pronounced ray-key) sounds mysterious, yet it has a straightforward premise: a simple, gentle transfer of energy that accelerates the body's ability to heal physical and mental disorders. Dr. Mikao Usui, a minister, scholar, and philosopher in Japan, brought Reiki's origins to light. In the mid-1800s—after a long search—Dr. Usui found a Sanskrit sutra written by a disciple of Buddha more than 2,500 years ago. He called his discovery Reiki and began what became the Usui System of Natural Healing. Reiki is not a spiritual belief system; rather it is used for self-healing and for offering healing to others. Many Reiki practitioners believe it is also a powerful way to help animals, supporting both conventional and holistic treatments.

How to Give Your Older Cat a Massage

Begin by paying attention to specific speeds, hand parts, and pressures. Your cat will patiently guide and encourage your progress.

Speed. Older cats especially enjoy long, slow loving caresses. Begin by stroking your cat down her back. Count how long that stroke takes. Caress again and double the time it takes to complete the stroke. What may seem excruciatingly slow to you feels good to your cat. Slowly explore every bump and contour. You can also try "no-motion" by allowing your cat's head to rest on a single finger or in your cupped palm.

Hand parts. Repeat the same technique with fingertips, finger pads, full palms, four fingers or two fingers together, and thumbs. Use knuckle nooks—a flat surface formed between your knuckles when you bend your fingers—for stroking under the chin, around the cheek, and behind the ears. Take time to discover which hand parts you and your cat prefer and favor.

Cat parts. Explore beyond the familiar regions of your cat's neck, back, and top of the head. Try caressing his furry chest, reach in for an underarm tickle, and slowly weave your fingers around his paws and claws.

Pressures. Frail older bones and bodies need gentler hand pressure. Your older cat prefers a mild pressure. Try using a light pressure (you'll just barely feel the muscles below the fur), or feather-light pressure (a fairy-light touch of the fur).

White gloves and brushes. For an additional massage treat, touch your cat while wearing a soft cotton glove. The different texture adds a refreshing change. Routine brushing, especially important for older cats, will feel better to your cat when combined with cat massage techniques.

Repetition. Don't subject your elderly cat to any sudden changes during a massage session. When you discover a technique that particularly impresses him, repeat it over and over again. For example, instead of stroking along the back two or three times, repeat this motion up to 10 times. Then you can alternate hand parts, speeds, and pressures. Consider devoting an entire massage session around a specific technique.

Friendly feline feedback. It's common for cats to become more affectionate and responsive with frequent massage. Some cats will even bypass the food bowl to demand affection. Others will knead. Pushy head-butts and cat kisses signify your cat's approval. With practice and patience, you may be serenaded with meows or treated to "power purrs."

—Courtesy of Maryjean Ballner, author of *Cat Massage* and *Dog Massage*

Flower Essences for Older Cats

Individual Essence	Present State	Desired State
Big Springs Rock Water	Vitality, seniors, indoor animals	Vitality, attunement to nature
Bleeding Heart	Consequences of a broken heart	Heart center opening, trust
Douglas Iris	World-weariness, lethargy, seniors	Enthusiasm for life
Douglas Violet	Withdrawnness in senior animals	Attunement to the present moment
Farewell to Spring	Depression due to aging	Peace of mind, fearlessness
Forget-Me-Not	Loneliness, isolation, accidents	Integration, acceptance, joy
Heartsease	Consequence of abuse, grief	Trust, heart opening
Mallow	Insecurity, anxiety, fear with age	Sense of divine love, protection
Marsh Marigold	Lack of deep attunement	Heart-centered communication
Pine Violet	Shyness, withdrawnness in seniors	Presence in the moment
Shasta White Lupine	World-weariness, senior years	Perseverance, stamina

—Reprinted with permission, Anaflora Individual Flower Essences

During a hands-on treatment, a Reiki practitioner rests his or her slightly cupped hands gently for three to five minutes or more in each of a number of positions corresponding to major organs, chakras (energy points), and endocrine glands of the body. During treatment positions, the fingers are held together and hands rest gently on the animal receiving Reiki. The process is said to work by balancing the energy of the body, then focusing on the injured area, tumor, or incision that may be present. Reiki can also bring comfort and peace to dying animals.

Each cat responds to Reiki in her own way. Some love Reiki and can't seem to get enough; others may back off or appear restless at first. Reiki masters say that animals will communicate what they want and need, often responding to Reiki more quickly than people do. One Reiki master explains that the elder cat is drawn to the energy that supports him. He recognizes and feels the energy, and intuitively knows it helps and heals.

TTOUCH

The Tellington-Touch, or TTouch, is a method based on circular movements of the fingers and hands over the body. Described like turning on the electric lights of the body, TTouch is designed to activate cell functions and awaken cellular intelligence, according to its inventor, internationally recognized animal expert Linda Tellington-Jones. TTouch can be used to assist with recovery from

Ear TTouch

The Ear TTouch, considered one of the most significant touches, is particularly effective for older cats. To start, gently but firmly stroke from the base of the ear to the tip. Supporting your cat's head with one hand, hold your thumb on the outside and folded forefinger on the underside of the ear with the other hand. Repeat this motion several times from the base of the ear to tip, covering different portions of the ear with each slide.

Next, create tiny circles between the thumb and forefinger, covering the entire ear in long lines. Finally, using the tips of the fingers make tiny circles, one at a time, all around the base of the ear.

illness or injury, including problems associated with aging, or to enhance the quality of an animal's life. It also helps establish a deeper rapport between people and companion animals. TTouch can be performed on the entire body, and each circular hand position or movement is considered complete within itself. Certified practitioners around the world teach TTouch techniques to those who want to share them with their companion animals.

AROMATHERAPY

Essential oils, or concentrated pure plant extracts, have been used to benefit the body, mind, and emotions since ancient times. Well-known for their therapeutic value and fragrance, the use of such oils is called aromatherapy.

Although scent therapy has been around for centuries, holistic veterinarians have only recently begun to incorporate its uses in their practices. They have discovered that certain aromas can have a powerful effect on animals through absorption into mucous membranes. For example, lavender can promote relaxation, while other scents can relieve anxiety or reduce blood pressure. Aromatherapy can be applied through diluted drops on the inside of the ear tip or released into the air through a diffuser. Check with a holistic practitioner before treating a pet on your own, though. Some oils, like pennyroyal, can be dangerous to cats.

ABCS OF CAT BODY LANGUAGE

People and cats who live together a long time understand each other well. A simple movement, the blink of an eye, or a single meow evoke secret conversations without words. Yet even those new to a cat's company quickly learn to appreciate the difference between an angry hiss and a cat kiss. Chances are your older cat reads you well. Can you say the same? The following sidebar can help you discern some of her more subtle body signals.

How to Read Your Cat	
Calm, relaxed cat:	Eyes—shut (contentment, bliss); stare, slow blink—called cat kiss (happy) Ears—perky, lean slightly forward Whiskers—straight out to sides Tail—soft twitching (comfortable); held high (happy)
Fearful cat:	Eyes—wide open Ears—flat against head Whiskers—flattened against face Tail—tucked low between legs
Curious cat:	Eyes—alert stare Ears—swivel Whiskers—fanning Tail—slightly curved over back (play invitation)
Angry cat:	Eyes—narrow, pupils dilate Ears—straight back Whiskers—angled forward, spread apart Tail—lashing, fluffed (back off)

CONNECT QUIETLY

Contemplation is your cat's natural state of being. To better connect and communicate with your senior cat, try sitting quietly with her, suggests Sharon Callahan, an internationally recognized animal communicator in Mount Shasta, California. Meditation, quiet inner listening, and centering prayer can help. Your cat will feel your love and concern, and you will feel and hear within yourself everything you need to know, Callahan says.

SECTION 4

AGE-RELATED HEALTH PROBLEMS

While many cats stay healthy throughout their lives, others will require the help of multiple disciplines—including internal medicine, clinical pathology, surgery, and nutrition— to heal injuries and solve health problems.

CHAPTER EIGHT

First-Aid Primer

No matter how carefully you protect your cat, accidents can and do happen. Cats love routine but don't always act predictably, thereby setting in motion the informal version of Murphy's Law, which states that if anything can go wrong, it will. A little forethought and some good planning can help derail this universal principle. Be prepared to treat common injuries your older cat may suffer until the cavalry arrives (or you reach the fort). Possessing the knowledge and tools to handle a feline emergency can ease a traumatic situation for both of you and, in addition, may save your cat's life.

How you deal with an emergency
can make all the difference in your older cat's survival.

WHAT DO I DO FIRST?

Contact your veterinarian in an emergency. While this advice seems obvious, it is easy to overlook when you feel panicky or frightened. Take a moment to telephone your vet when your cat is injured or suddenly appears seriously ill. A veterinary professional can offer advice on how to proceed when you may not be thinking clearly. Your call also paves the way for veterinary hospital staff that will anticipate and prepare for your cat's arrival.

Of course, most veterinarians are not available around the clock. You can save precious minutes by knowing your doctor's plan for handling emergency situations ahead of time, especially those occurring after normal business hours and on weekends (when emergencies always seem to happen—see Murphy's Law above). Keep the telephone numbers of your veterinarian, a local emergency veterinary service, local animal poison center, and the national ASPCA Animal Poison Control Center handy at all times.

While getting your cat to your veterinarian quickly is the goal, an unexpected feline accident or injury nevertheless calls for quick, and often challenging, action on your part. For example, you may not have time to get professional help for a choking

cat; you must act immediately to save her life. Healthy senior cats can spin into life-threatening shock in an emergency; a debilitated older cat, however, already is in a fragile state. Unless you are a doctor you cannot be expected to act like one. Nevertheless, how you deal with an emergency can make all the difference in your older cat's survival. Studying the basics of first aid can help. First-aid measures for common emergencies you and your cat may encounter follow.

FIRST STEPS

1. Remain calm. Your adrenalin may be running full-tilt, but you need to approach your injured cat quietly and in a nonthreatening way. Take a few moments to assess the situation.

2. Do not lift an injured cat with your bare hands. Pain can cause even the tamest cat to lash out and act defensively.

3. Use a large towel, folded blanket, or small rug to carry your cat to a veterinary hospital.

4. Cover your cat with an emergency "space blanket" or bubble wrap (a jacket or blanket will do) to conserve her body heat and help prevent shock, which occurs when the body cannot maintain sufficient blood pressure.

SHOCK

Shock is a medical emergency that can occur when a cat is severely hurt or ill—especially if she becomes unconscious—and is potentially fatal. Signs include pale or blue lips and gums; a faint but rapid pulse; shallow breathing; staring, unresponsive eyes; a cold body; and lethargy. A cat in shock must be kept warm, as specified above. Quick veterinary care is critical.

BLEEDING

What to do: Apply a pressure bandage to any wound that causes severe external bleeding and rush your cat to the veterinarian. (A minor cut or scratch is likely to stop bleeding on its own and will not bleed profusely unless a larger blood vessel underneath opens.) Apply a sterile gauze pad or clean cloth directly over the wound. Use adhesive or masking tape to secure the bandaged area, or firmly tie the bandage in place with gauze or cloth strips. Unless directed by your veterinarian as a last resort, do not use a tourniquet, which can cut off the blood supply and cause additional damage to the wounded area. Don't

remove an object (such as a bullet) from a wound. Don't lift the bandage to see if the bleeding has stopped. Do take your cat to the veterinarian for additional care.

CHOKING

What to do: The usual suspect is an object lodged in the mouth or throat. If your cat is conscious, proceed cautiously to avoid being bitten. A panicked cat won't appreciate your help. Wrap him in a towel or other restraint and, if available, ask someone else to hold him firmly. Open your cat's mouth by firmly pressing fingers on either side of his jaw. Look inside to determine what's wrong before sticking your fingers in his mouth—you don't want to push a foreign object deeper into the throat. If you can reach the offending item, pull gently with your fingers, tweezers, or long-nosed pliers. Follow the same procedure for an unconscious cat. Caution: If you encounter resistance, do not try to force the object out. If you discover your cat has swallowed string or yarn, do not attempt to remove it. Deeply lodged string or thread can cut through the walls of the intestines,

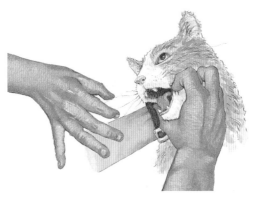

> ## Veterinary Tip
>
> To clean a minor cut more easily, dip a blunt-tipped pair of scissors in tepid water, then gently clip the fur around the wound. The clipped fur sticks to the blades, not the wound, and is easily removed by rinsing the scissors in a cup of water.

or may have a needle attached to it. Instead, rush him to a veterinarian.

Can't reach the blockage? Try a modified Heimlich maneuver to loosen the object. To do this:

1) Place your cat on her side.
2) Position the heel of your hand behind the last rib, angling slightly up. Apply three or four quick thrusts with gentle but firm pressure (you don't want to crack ribs).
3) If the cat does not "cough" out the object, repeat the process. If still unsuccessful, you will need speedy veterinary help.

FRACTURES AND DISLOCATIONS

In an open or compound fracture, the bone breaks and punctures the skin. A simple fracture, which doesn't pierce the skin, usually reveals itself through swelling, painful movement, or a dragging limb (also a symptom of dislocation). A lower jaw that hangs open may signal a broken jaw.

What to do: If your cat appears to have a fractured limb or dislocated hip, carefully place him on a large, flat surface. A cardboard box or clean litter pan works well. Cushion the "gurney" with a towel or blanket and keep your cat as still as possible to avoid further injury. Cover your cat to keep him warm and prevent shock. Transport him to your veterinarian immediately. Leave splinting a fractured limb to a professional, since this process can cause severe pain. If you suspect a fractured jaw, tie gauze tape or a long piece of cloth under the chin and behind the ears.

BURNS

What to do: Prompt veterinary treatment is needed to treat major or minor burns on a cat's body. For minor burns, cover the burned area with cool compresses (don't use ice!) on your way to the veterinary hospital.

For major burns, protect the area with a thick layer of gauze or cloth, cover your cat with a blanket, and seek immediate veterinary help. Protect your hands with rubber gloves if you suspect chemicals caused the burn.

Do not apply antiseptic ointments, butter, or other products to any burned area unless your veterinarian advises you to do so. Never cover any type of burn with cotton balls or pads, because cotton fibers will stick to the affected area.

FROSTBITE

Older cats are more prone to frostbite than younger cats. The extremities are the most likely body parts to be affected; the pads of the feet, tips of the ears, and the tail are the most vulnerable areas. Look for pale, reddened skin that becomes hot and painful when touched. Swelling also may occur, along with hair loss and peeling skin.

What to do: Place the cat in a warmed area and thaw the frostbitten areas slowly. To do this, use warm, moist towels and replace them frequently. Stop the procedure when the affected tissues become flushed. Wrap your cat in a blanket to conserve her body heat. Warning: Do not apply anything hot to frostbitten skin, and do not rub or massage frozen tissues—it may cause more damage. Consult your veterinarian before applying any topical ointment, and keep your cat out of the cold in the future, as frostbitten tissues are more vulnerable to repeated freezing.

POISONING

The clinical signs of poisoning may be as dramatic and terrifying as a seizure, or as simple as drooling. Symptoms vary from no initial signs at all to staggering, weakness, lack of coordination, depression, vomiting, diarrhea, and convulsion, among others.

What to do: Don't panic if you suspect your cat is exposed to a poisonous substance. Take a minute to collect the material involved, including a product container or anything your cat has chewed or vomited. Put it in a zip-lock bag. If your cat loses consciousness, seizures, or has difficulty breathing, seek immediate veterinary help.

If you are unable to reach your veterinarian or a local animal poison-control center, contact the ASPCA Animal Poison Control Center, an operat-

ing division of the American Society for the Prevention of Cruelty to Animals. The Center hotline provides 24-hour coverage.

888-4ANI-HELP

(888) 426-4435 or (900) 680-0000

Before you call, be ready to provide your name, address, and telephone number; information concerning the exposure (what poison, the amount, and time elapsed since exposure); the age, sex, breed, and weight of your cat; and the symptoms your cat exhibits. Note: The nonprofit Center charges a moderate professional fee per case.

Recipes for Poison

Toxic chemicals, dangerous plants, products, and substances found in our daily environment can prove poisonous or fatal to cats of any age. Watch for the following common but potentially dangerous elements.

Flowers
Lilies are lovely, but several types are highly poisonous to cats, including the Easter and tiger lilies. Ingesting even a small amount can cause kidney failure in cats.

Household plants
Azalea, dieffenbachia, holly, hyacinth, oleander, philodendron, poinsettia, and sweet pea are among common indoor and outdoor plants toxic to cats. For a more complete list, visit the ASPCA Animal Poison Control Center Web site at www.napcc.aspca.org.

Flea products

Never use insecticides on elderly or debilitated cats without first consulting your veterinarian. Use a vet-recommended flea product, or consider using a nontoxic, hands-on method, such as removing fleas with a flea comb and then submerging the fleas in a small container of soapy water (a time-consuming but much safer alternative). Always read the label before using any product on your cat. A label that reads "do not use on cats" should never be ignored; even a small amount of some canine flea products can be fatal to cats. For example, cats are highly sensitive to permethrin, a common ingredient in flea-control products, and can experience tremors and seizures if only a few drops are applied. (Note: Some products specifically labeled for use on cats do contain tiny amounts of permethrin, usually less than 0.1%. When used according to label instructions, however, these products can be used safely on cats.)

Medicine mix-ups

Caretakers with older cats on prescribed medications sometimes give human medications to their cat unintentionally—a common occurrence in the morning when sleepy people reach for the wrong container. Even something as ordinary as a pain reliever containing acetaminophen (found in Tylenol and similar products) can kill a cat. Keep your medications separate from your cat's pills, and put kitty stickers or bright, colored stickers on the medications to avoid confusion.

Dangerous foods

Don't let your cat ingest moldy or spoiled foods, onions, onion powder, alcoholic beverages, or chocolate.

Auto/garage items

As little as one teaspoon of sweet-tasting conventional antifreeze (which contains ethylene glycol) can be deadly to cats. For a safer alternative, choose the less-toxic version of antifreeze containing propylene glycol.

Household cleaning supplies

Watch out for bleach, ammonia, cleaning fluids, disinfectants, and drain cleaners. While your older cat isn't likely to drink such things, she may haphazardly walk through a product in use, and ingest it as she licks the substance off her paws or fur. Keep the toilet lid on as well, especially if the water is chemically treated.

Gases

Inhaling gases or fumes from ammonia, carbon monoxide, and cooking gas can be toxic to cats, who may show signs of inhalation poisoning with dizziness, difficulty breathing, and bright red lips and tongue. Fresh air is the antidote; then seek veterinary help.

In addition to your pet safety kit (see sidebar, page 90), a well-stocked first-aid kit goes a long way toward protecting your cat during an unexpected crisis. Pre-packaged kits can be purchased at your local pet supply store, or you can create your own. It doesn't take much to build a first-rate first-aid kit for your cat (see page 95).

BREATHING STOPS, NO HEARTBEAT

A cat's basic life functions can cease when she becomes unconscious due to a traumatic injury, accident, or illness. When a cat ceases breathing and her heart stops, a small window of roughly

Pet Safety Kit

Be prepared! Keep a pet safety kit on hand for poisoning emergencies. Place the kit in an easy-to-remember, accessible area and include the following items:

- A fresh bottle of 3% hydrogen peroxide (USP)

- Can of soft cat food

- Turkey baster, bulb syringe, or large medicine syringe

- Saline eye solution to flush out eye contaminants

- Artificial tear gel to lubricate eyes after flushing

- Mild, grease-cutting dishwashing liquid to bathe cat after skin contamination

- Rubber gloves to prevent exposure to skin contaminants

- Forceps to remove stingers

- Restraint (large towel or folded blanket) to keep a cat from hurting you while it is excited or in pain

- Pet carrier to help carry your cat to your veterinarian

—Reprinted with permission,
ASPCA Animal Poison Control Center

three to four minutes exists before permanent brain damage occurs.

Caution: Do not attempt artificial respiration or cardiopulmonary resuscitation (CPR)—a lifesaving procedure combining artificial breathing and chest compressions—on a cat who is conscious and breathing! CPR is a stopgap method to keep your cat alive until her heart and lungs can resume their work; perform CPR only when a cat's heart and breathing have stopped.

What to do: To breathe life back into your cat, think ABC—

- A for Airway

- B for Breathing

- C for Circulation

Is your cat breathing? To decide if your cat requires artificial respiration, check for airflow near the mouth and nose, and observe whether your cat's chest rises and falls. A blue tongue indicates a lack of oxygen. If your cat is not breathing, place her on her side on a flat surface and remove her collar.

A for airway: Remove any obstruction from the mouth or throat (see Choking, page 85). Open her mouth and gently pull her tongue forward to avoid blocking the airway. Gently close her mouth, and pull her head and neck forward. To express old air from her body, gently place hands on ribs and push down quickly. Release. Your cat's elastic recoil mechanism will refill her lungs with air. Repeat approximately once every five seconds.
If the recoil mechanism fails to work (due, for example, to a chest wound or collapsed lungs), you will need to blow air into your cat's lungs.

B for breathing: To perform artificial respiration, support your cat's head in your hands, keeping her

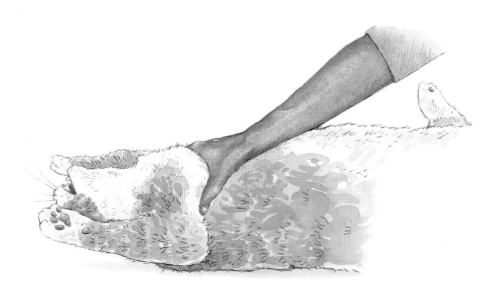

mouth closed and her tongue pulled forward. Place your mouth over her nose and gently but consistently blow into your cat's nose for about three seconds. Her chest will expand; pause to release the air. Repeat once every four to five seconds, or up to fifteen times per minute, until your cat breathes on her own. Veterinary care is a must.

C for circulation: The beating of your cat's heart can be felt by placing your hand (or ear) on the left side of her chest above and behind her elbow, or by pressing arteries in the neck or inside upper hind leg. No pulse and an absence of breathing equal cardiac arrest, and you'll need to begin CPR immediately.

CPR Steps

1. Place the palm of your hand on the cat's chest; rest your thumb on his left side at the point of his elbow while your fingers lay flat on your cat's right side for compression. (See illustration above.)

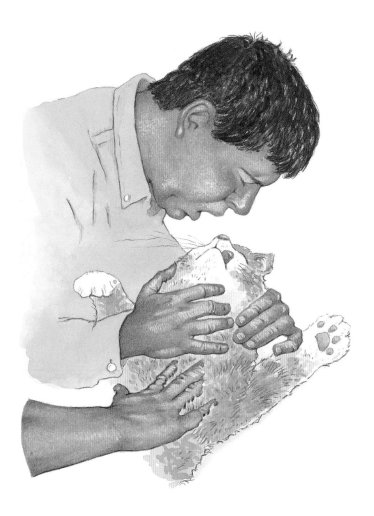

2. Squeeze your hand smoothly (gently but
   firmly) approximately once every second.
   Repeat chest compressions five times, then
   follow with one breath of artificial respiration. (If someone is available to help, divide
   the job.)

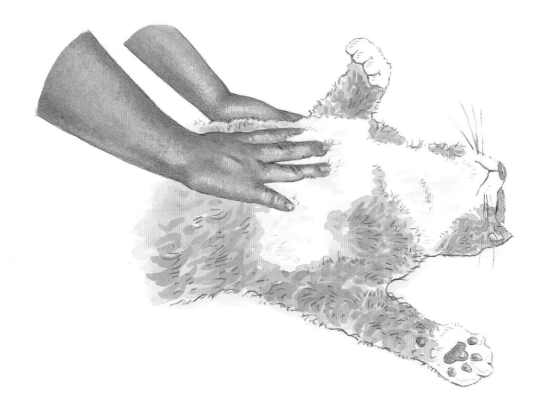

3. Continue this rhythmic sequence until your cat's heart begins to beat or until a veterinarian can be reached. Be sure to check for a pulse frequently while performing CPR.

Last Word on CPR

Keep the following in mind: You don't have to be an expert, and you don't have to perform CPR flawlessly to save your cat's life. Your cat will not die because you fumble through CPR. A cat without a pulse or respiration will surely die, however, if no attempt is made to save his life. One veterinarian in emergency practice for a decade and a half found that numerous caretakers were able to revive their pets by gently blowing into the nose and massaging the chest. Many weren't well versed in how to use CPR. The point is to give your cat a small chance to survive in a situation that would otherwise prove fatal. Enroll in a pet first-aid class to learn more and gain some hands-on experience (and confidence) in pet CPR. To find a class in your area, contact your local humane society or Red Cross chapter.

Normal Feline Vital Signs
(General Guidelines)

Temperature: 100° F to 102.5°F
Breathing rate (respiration):
20 to 30 per minute
Pulse: 160 to 240 beats per minute

First-Aid Kit

Essential Items
> Nonstick sterile gauze pads (various sizes)
> Gauze elastic tape or adhesive tape to secure bandage
> 3% hydrogen peroxide (to clean wounds)
> Antibacterial ointment
> Cotton swabs, cotton balls
> Rubbing alcohol
> Rectal thermometer
> K-Y Jelly (to lubricate thermometer)
> Blunt-tipped scissors
> Tweezers
> Eyedropper
> Plastic syringe for liquid medications

Useful to Have on Hand
> Nylon cat muzzle, towel (for restraint)
> Hairball remedy (paraffin oil)
> Kaopectate (for diarrhea—ask your veterinarian for correct dosage)
> Styptic pencil or powder (for bleeding claws); —note: It stings!
> Activated charcoal capsules (for poisonings; check with vet before using,
>   as its use may be contraindicated)
> Antihistamine cream/lotion (for insect stings—avoid products containing
>   local anesthetics, which can cause methemoglobinemia, a blood disorder, in cats)

Pack the items in a small toolbox or other easy-to-carry storage container, and tape the telephone numbers of your veterinarian, emergency pet clinic, and poison control center to the top of the box. If you aren't sure how to use every item in your first-aid kit, ask your veterinarian for instructions before an emergency strikes.

Extra Rx Tips for Cats
> Consult a veterinarian before giving any topical or oral medications to your cat
> Induce vomiting only with veterinary approval
> Avoid using syrup of ipecac unless your veterinarian specifies its use—
>   an excessive amount or repeated doses may have a cardio-toxic effect on cats

CHAPTER NINE

Solutions for Common Health and Behavior Problems

More veterinarians now incorporate senior care into their practices. A mix of preventive and therapeutic strategies, senior care is based on the recognition that the needs of older pets are different than those of younger animals and should be addressed—and treated—accordingly. Senior care for cats broadens the medical, nutritional, and overall wellness programs established during the early years of their lives.

Early intervention can often
enhance and extend the lives of our cats.

While aging is not a disease, age-related problems are inevitable and generally irreversible. The good news is that early intervention can often enhance and extend the lives of our cats. This new direction in senior care emphasizes healthful living as pets age, not simply the treatment of old, debilitated patients. This means more frequent veterinary exams, diagnostic testing, nutrition counseling, and, most likely, an increased need for age-related medicines. The veterinarian can do none of the above, however, without your participation.

Just as the guardian of a new kitten gradually learns how to properly care for a cat, it's important to pay attention to your cat's needs as she ages. You will need to regularly work with your vet to deal with common health issues that crop up over the years. Catch them early and, above all, do not chalk up the following health and behavior problems to old age. Ignoring them can cause your cat pain and severely reduce your cat's quality of life. The ultimate "solutions" are in your veterinarian's hands, with your help. Working together is the key to your senior cat's health.

ARTHRITIS

Arthritis, defined as the inflammation of the joint, is not nearly as common in cats as in dogs, but it is seen most often in older cats. Osteoarthritis, or degenerative arthritis, can affect one or more joints, making it painful for a cat to move or walk. Traumatic arthritis occurs as result of injury to

joints—a cat who has been struck by a car is a prime candidate for this type of arthritis.

Signs: Legs appear stiff and joints may be warm to the touch. Cats may display reduced physical activity, lameness, difficulty rising and walking, and changes in posture.

Treatment: Veterinary-approved medications are given to reduce inflammation and relieve pain. Surgery may be advised in some cases.

What you can do: Pay attention to early signs of arthritis and seek veterinary help promptly. If your cat is willing, gentle massage may relieve pain and loosen tight muscles (see Massage on page 72). Wrap a towel around a hot-water bottle and apply heat to painful areas. Watch your cat's weight to relieve stress and strain on affected joints. Caution: Do not treat an arthritic cat with human pain relievers of any kind; they can be harmful or fatal to cats.

DENTAL DIFFICULTIES

Dental disorders, or periodontal disease, affect teeth and gums. If ignored, dental problems can eventually spread infection, speed disease, and damage major organs throughout a cat's body.

Signs: Drooling, bad breath, inflamed gums, loss of appetite and drop in weight, and tartar on teeth. Cats with painful mouth conditions may chew on one side, drop food, or display chattering teeth. Advanced dental problems include mouth ulcers and missing crowns on teeth.

Treatment: Veterinary care includes thorough cleaning under general anesthesia. Decayed or broken teeth are usually extracted, and antibiotics may be given to fight infection.

What you can do: Ask your veterinarian or a veterinary nutritionist whether dietary changes are in order. Seek regular dental check-ups and teeth cleaning, preferably twice a year (See Dental Care, page 57).

LITTER BOX BLUES

An older cat usually has a good reason for urinating or defecating outside the litter box, though humans don't tend to agree. Nothing erodes the relationship between cat and human faster than this problem. A visit to the veterinarian is imperative to rule out possible medical problems, such as cystitis, lower urinary tract disease, diabetes, or kidney disease, and to determine the cause of the problem. If diagnostic testing rules out a medical cause, a behavioral investigation is the next step.

INCONTINENCE

Older cats gradually lose muscle tone, a common cause of incontinence. Painful arthritis may deter an elderly cat from getting up and using the litter box.

Signs: Has frequent accidents, dribbles urine, and defecates while resting or sleeping.

Treatment: Veterinary-approved drugs. Be sure to ask about possible side effects.

What you can do: Seek veterinary help early. Ensure litter box is nearby, clean, and easily accessible.

How to Take Your Cat's Temperature

A fever is part of the body's natural healing process, but it may also indicate a serious health problem. Fever is especially worrisome in fragile older cats that dehydrate easily. A cat's normal temperature ranges from 100°F to 102.5°F.

To take your cat's temperature, lubricate the end of a rectal thermometer with K-Y Jelly and, if possible, ask a family member or friend to help hold your cat steady. Lift your cat's tail and gently rotate the thermometer into the rectum no more than 1 inch deep for one to two minutes. (An electronic thermometer can perform the job quicker, emitting beeps when your mission is accomplished and providing a clear digital reading.) A fever of 103.5°F or higher warrants a call to your veterinarian right away.

How to Give Your Cat a Pill or Liquid Medications

There's a reason so many stories describe the art and misadventures of giving a cat a pill. Cleveland Amory devoted a whole chapter to the subject of attempting to pill his cat Polar Bear in his international best-seller, *The Cat Who Came for Christmas*. Indeed, cats often seem to be masters at resisting pills. Some go so far as to spit a pill out as soon as the pill-giver smugly leaves the room.

The older your cat becomes, the greater your chance of medicating him grows. While this often comes under the grin-and-bear-it category, pilling your cat doesn't have to be an exercise in frustration or futility. You just need to figure out what works best for both of you. Here are some options.

- Wrap your cat firmly in a towel with only her head exposed. Kneel on the floor, position your cat between your knees, and hold your cat's head away from you, or have someone else hold your cat while you dispense the pill. With the pill in one hand, use the fingers of your other hand to gently pry open your cat's jaws (try the thumb and first finger). Put the pill as far back on the tongue as you can manage, then close her mouth. Keep it closed while you stroke her throat to make sure the medicine goes down. Praise her or give her a treat.

- Package the pill in something that entices your cat to eat it without a second thought. Try a soft cat treat, piece of meat, or cream cheese. Check with your veterinarian for an appropriate food choice for your cat.

- A pill gun, available from veterinarians, can make the pill-popping process easier.

- Turn your pill into liquid medicine by crushing it and mixing with water. To give your cat this solution, or any other liquid medications, fill a medicine dropper or dosing syringe with the correct amount. Restrain your cat as outlined above. Open his mouth, and put the end of the dropper into the corner of the mouth, or cheek pouch. Close the mouth around the dropper and quickly empty the contents. Your cat's natural swallowing reflex will take over.

CONSTIPATION

Constipation, though not likely to be life-threatening at first onset, should not be left untreated. This condition can signal a serious underlying, or developing, medical problem. Causes may involve weak muscles, lack of exercise, stress, and poor diet.

Signs: Passes dry stools, crouches and strains while attempting to defecate. May also express soft stools with blood or mucus, and lose appetite and weight.

Treatment: Laxatives and a change to a higher-fiber diet often solve the problem. Surgery may be needed to remove internal blockage. Fluids may be given for dehydration.

What you can do: Seek veterinary help early. Follow veterinary recommendations; in many cases, this means a diet high in fiber and feeding an overall balanced diet.

PERSONALITY CHANGES

Older cats sleep more. They may become less socially inclined and crankier. Sudden or unusual behavior changes, however, are a sign that something is wrong and should be brought to the attention of your veterinarian. For example, a cat that suddenly becomes aggressive may be suffering from a painful dental problem. Other changes may include depression, agitation, lethargy, and Alzheimer's disease-like behavior.

CDS

CDS, or cognitive dysfunction syndrome, is common in elderly cats. A progressive, age-related disease, CDS is caused by physical and chemical changes that affect brain function. Signs of CDS include altered sleep cycles, increased vocalization, confusion, aimless wandering, excessive grooming, aggression, and inappropriate elimination. In addition, a phobia, such as a sudden fear of the dark, may appear. The clinical signs relate to senility or impaired mental function. CDS symptoms can be managed by drug treatments, exercise, and oxygen therapy after other health conditions or diseases are ruled out. A desensitization program can help you deal with your cat's fears. Your veterinarian can help you devise the right course of action. In addition, treat your cat with compassion, understanding, and loving care.

CHAPTER TEN

Illnesses Your Older Cat May Develop

Aging increases the odds of contracting a debilitating illness or disease in both people and cats. Our respective bodies simply begin to wear down with time. Because cats age more quickly than we do, it's logical to deduce that feline health problems can develop faster as well. Of course, genetics, environment, nutrition, and preventive care—or the lack of these—all play a part in the way your cat ages. While many cats stay healthy throughout their lives, others will require the help of multiple disciplines—including internal medicine, clinical pathology, surgery, radiology, and nutrition—to maintain health and a good quality of life.

*Early recognition, combined with better diagnostic aids
and treatments, are helping more cats live longer and better.*

Fortunately, cats diagnosed with a serious disease have more treatment options than ever before. Today, new technology is available at the touch of a keyboard to help veterinarians gather expert advice and diagnose illnesses. Despite technical advances, the diagnosis of a serious disease can seem frightening and final. It's important to keep in mind that early recognition, combined with better diagnostic aids and treatments, are helping more cats live longer and better. Thus a diagnosis of cancer is no longer a death sentence but a health challenge that may be contained, treated and, in many cases, cured.

Note: Symptoms of feline disease often have common denominators such as appetite loss, increased thirst, or frequent urination. Do not try to diagnose a health problem yourself. Consult a veterinarian if you suspect your older cat may be ill.

THE BIG FOUR

The four leading causes of death in older cats are renal failure (kidney disease), cardiovascular disease, cancer, and diabetes mellitus. The best defense remains a good offense, however, so let's take a closer look at illnesses your cat may develop.

Kidney Disease
Your cat's kidneys filter waste products from his body; regulate electrolytes; produce a hormone,

erythropoietin, which helps stimulate bone marrow to produce red blood cells; and produce renin, an enzyme that helps regulate blood pressure. Also known as chronic renal failure (CRF) and chronic renal insufficiency (CRI), kidney disease is a progressive and irreversible deterioration of kidney function. Early signs are subtle and easily missed until roughly 70 percent of the kidney no longer functions.

Signs: Increasingly thirsty, excessive urination, loss of appetite, nausea and vomiting, weight loss, decrease in activity, anemia, hypertension, poor hair coat, emaciation, and dehydration.

Cause: Typically age-related, with genetics, environment, and disease as contributing factors.

Treatment: No cure, but manageable with extra fluids, medications, and special diets. With a strong commitment from caregivers, cats with CRF can live for months to years with a decent quality of life. Kidney transplants are an accepted and relatively safe treatment. The success rate for candidates in good condition ranges between 80 and 90 percent.

Cardiovascular Disease
The cardiovascular system is, pun intended, the heart of your cat's body. The ability of the heart to pump blood efficiently decreases with age. This process

AGE-RELATED HEALTH PROBLEMS 107

Signs: Heart and circulatory disorders include labored breathing or panting, lethargy, loss of appetite, rear leg pain or paralysis.

Treatment: Diuretics to remove excess fluid, lifelong drug therapy, dietary management, and possible taurine supplements.

Cancer

Cancer, or neoplasia, is a term covering a group of diseases. Cancer develops when abnormal, or uncontrolled, cell growth replaces normal tissues and disrupts body functions. The growths, or tumors, may be benign or malignant. The care of feline cancer patients is becoming a major component of more veterinary practices as the feline pet population increases and grays. Advances in feline oncology have increased survival times, improved treatment response, and prolonged disease-free intervals. Many cats with cancer can be cured or rendered free of disease for significant periods of time.

Signs: See sidebar on next page.

Treatment: Surgery (to remove tumor), chemotherapy, and radiation therapy. Magnetic resonance imaging (MRI) improves diagnosis. Immuno-augmentive therapy is used in some cases.

Diabetes Mellitus

Diabetes mellitus is a disease in which a cat's body cannot control her blood sugar levels. A common disorder of the endocrine system, diabetes mellitus is caused by a lack of insulin, a hormone that stimulates the movement of glucose (sugar) from the blood into cells. The pancreas

makes the system work harder, which can cause heart muscles to weaken. Congestive heart failure is the result, occurring when one of the heart's lower ventricles, or chambers, can no longer do its job.

Cardiomyopathy, a disease of the heart muscle, is the most common heart problem in cats. Hypertrophic cardiomyopathy (HCM), the most frequently diagnosed version, causes portions of the heart to thicken and pump blood less efficiently. While HCM prevents the heart from filling with blood adequately, dilated cardiomyopathy (DCM) affects the heart muscle's ability to contract and thus pump blood. In DCM, the muscle walls thin and the heart enlarges. The rarer restrictive cardiomyopathy (RCM), also known as intergrade or intermediate cardiomyopathy, interferes with both the filling and pumping of the heart.

produces insulin; diabetes mellitus occurs when the pancreas fails to produce enough of it. Diabetes mellitus is most commonly seen in senior cats. Unspayed female cats are at increased risk, as are overweight cats of both sexes.

Signs: Increasingly thirsty and hungry but losing weight, increased urination.

Treatment: Stabilize with fluids and medications. Oral medications for a lucky few, daily insulin shots for most, along with dietary management.

Hyperthyroidism
A common disease of the feline endocrine system in older cats, hyperthyroidism is associated with overactivity of the thyroid gland. Hyperthyroidism may mask underlying renal, or kidney, disease.

Signs: Increased drinking and ravenous appetite, combined with weight loss, vomiting, diarrhea, hyperactivity, hypertension, and rapid nail growth.

Treatment: Lifelong medication, radioactive iodine therapy (works by destroying the thyroid gland), or surgery (removes affected thyroid lobes).

Inflammatory Bowel Disease (IBD)
A controllable disorder (especially if caught early) of the small intestine, IBD commonly occurs in older cats. Your veterinarian must diagnose this condition after ruling out other causes of gastro-intestinal disease.

Signs: Anorexia or weight loss, frequent vomiting, and diarrhea.

The Veterinary Cancer Society's Common Signs of Cancer in Animals

Early cancer detection is critical, because screening tests for specific types of cancer are not yet available in veterinary medicine. Do not attribute the following warning signs to old age.

1. Abnormal swellings that persist or continue to grow
2. Sores that do not heal
3. Weight loss
4. Loss of appetite
5. Bleeding or discharge from any body opening
6. Offensive odor
7. Difficulty eating or swallowing
8. Hesitation to exercise or loss of stamina
9. Persistent lameness or stiffness
10. Difficulty breathing, urinating, or defecating

Treatment: Dietary management, intestinal antibiotic, or anti-inflammatory drug.

Liver Disease
Liver disease refers to any condition that interferes with the liver's normal function. This includes hepatic lipidosis, a common and potentially fatal liver ailment primarily affecting overweight cats.

Signs: Refusal to eat, increased drinking and urination, jaundice, abdominal fluid build-up, vomiting, diarrhea, depression, and lethargy.

Giving Injections

Most cat caretakers are a little nervous about giving a cat an injection at first. Don't worry! With practice, you'll soon be able to fill a syringe and inject your cat like a pro. You may be called upon to give injections subcutaneously (meaning under the skin) regularly if your cat is diagnosed with diabetes (insulin injections) or kidney disease (fluid injections under the skin). Your veterinarian will explain the process and help you get started. She will explain that subcutaneous injections are usually placed under the loose skin at the back of the neck or between your cat's shoulder blades. Practice until you feel confident you can handle the procedure. You will soon develop a routine that is comfortable for both of you. Extending your cat's life and maintaining her quality of life are worth overcoming any qualms about giving injections. Here are a few tips.

- Choose a consistent, comfortable room or area in which to give injections.

- Don't rush. Take time to sit quietly and bond with your cat.

- Furnish a healthy treat or a meal after treatment. This will help your cat associate the injection with something pleasant.

Treatment: Immediate veterinary care is needed; force-feeding, then special diet to reduce the need for liver function.

SHOULD I VACCINATE MY OLDER CAT?

Deadly feline viruses can strike cats at any age, and vaccines play a crucial role in controlling these infectious diseases. Immune function declines as cats age, which, in turn, increases a senior cat's vulnerability to infection. Older cats, for example, run a higher risk of developing respiratory disease, which vaccinations can help protect against. Regular booster vaccinations are still recommended for healthy older cats, but vaccination protocols have changed.

Interest in vaccine-related issues has increased, thanks in large part to safety concerns (such as a chance of developing a life-threatening sarcoma— a type of tumor—at the vaccination site) and questions regarding the span of immunity vaccinations provide. For example, one anti-immunization argument reasons that people do not need yearly vaccinations, so why do pets require annual booster shots? These issues will continue to be addressed by national veterinary organizations, scientists, and vaccine manufacturers. According to the latest American Association of Feline Practitioners/Academy of Feline Medicine (AAFP/AFM) Feline Vaccination Guidelines, vaccinations for older cats should be administered based on individual risk assessment and in compliance with local laws. In general, debilitated or ill cats should not be vaccinated, while healthy

seniors can be vaccinated in the same manner as younger cats.

After receiving the initial vaccination series, and a booster one year after the primary vaccination, healthy senior cats should be revaccinated no more frequently than every three years for the following:

FPV: Feline parvovirus, cause of feline pan-leukopenia
FHV: Feline herpesvirus-1, cause of feline viral rhinotracheitis
FCV: Feline calicivirus
RV: Rabies virus

Several other vaccines exist and are generally considered optional, depending on factors such as age, risk, and the likelihood of infection or disease if exposed. For example, the Feline Leukemia Virus (FeLV) infects domestic cats throughout the world, but outdoor cats that come in contact with other infected cats are particularly vulnerable. The FeLV vaccination, however, is not recommended for older indoor cats with a minimal risk of exposure to FeLV-infected cats.

Health Insurance for Cats?

Veterinary care is fresh on the heels of human medicine in terms of technology and development. While veterinary costs haven't escalated as rapidly as have human health-care costs, a chronic feline health problem or disease can quickly drain a wallet. No one wants to be forced to choose between a beloved cat's life and money.

To alleviate the shortfall, a growing number of cat owners carry pet health insurance, similar to the health insurance they use to help shoulder their own medical bills. Several companies now offer some form of pet health insurance. Veterinary Pet Insurance (VPI) in Anaheim, California, was the first, pioneering the concept in 1980 with the support of 750 independent veterinarians. Today VPI policies are licensed in every state and the District of Columbia. The company offers several plans covering everything from cancer treatments to homeopathic medicine, vaccinations, and other routine care. Endorsed by the American Humane Association, VPI has issued more than one million policies and enjoys an 82 percent renewal rate.

VPI recently broadened its coverage to provide additional benefits for cats. Pet owners in 41 states can choose the VPI Superior and VPI Standard Plans featuring lower rates for cats and increased surgical benefits, along with enhanced benefits for specialized diagnostics, such as MRIs and CAT scans. Annual benefit maximums range from $9,000 to $14,000, with monthly rates beginning under $10 for cats.

For more information, call (800) USA-PETS (800-872-7387) or visit www.petinsurance.com

THE LAST DANCE

Older cats teach us patience and courage,
how to let go gracefully and, finally,
how to heal and begin again.

CHAPTER ELEVEN

Coping with Chronic Disease and Pain

Most of us do not think about caring for an ill or dying pet when we bring a young cat into our heart and home. We may not stop to consider that the adoption contract, written or unwritten, is a lifelong commitment to see our cats through illness, injury, or simply old age. Thanks to the bond we share with our cats, though, most of us make this commitment willingly, no matter how hard it may prove to be. Older cats, like other valued family members, deserve nothing less than the best care we can provide during the last stages of their lives.

Cats feel pain in much the same way that we do.
The difference lies in how each species responds to pain.

How Can I Tell if My Cat Is in Pain?

The following signs can signal discomfort:

• Hiding

• Reduced activity, plays less

• Failure to use a litter box

• Obsessive grooming

• Lameness or reluctance to use a limb

• Lethargy

• Loss of appetite, failure to eat

• Drooling, panting

• Restless, agitated

• Flinches or trembles

• Becomes vocal or cries, may hiss or strike out

• Suddenly irritable or aggressive

Caring for a chronically ill cat can be a struggle and a gift, both physically and emotionally. You may not feel up to the challenge of caring for a seriously ill cat, yet it's important to realize that you are the most qualified person to do so. The loving connection established long ago often grows as life winds down. We realize, more than ever, how precious each moment has become.

PAIN MANAGEMENT

Cats are wired similarly to humans from both a physiological and neuroanatomical standpoint. This means cats feel pain in much the same way that we do. The difference lies in how each species responds to pain. We communicate our discomfort to others, seeking help and relief quickly. Cats, on the other hand, take a different approach. They prefer to hide when they hurt, a carryover from behaviors developed to protect themselves from predators.

THE FOURTH VITAL SIGN

In its simplest terms, pain is stress. Chronic or severe pain significantly stresses the physical body and affects emotions. Pain can impair wound healing and slow recovery. For people and cats, stress can lead to metabolic changes that effect the immune system, which, in turn, can lead to potentially life-threatening complications. Pain can, in a sense, kill a cat.

A cat may display a poker face when it comes to hiding pain, but the understanding of human pain management has grown over the past decade, and veterinary medicine has followed close behind. Veterinarians on the cutting edge of pain management now rate pain as the fourth vital sign in animals, along with temperature, respiration, and blood pressure.

For veterinarians, whose patients do not walk into a hospital or clinic on their own, the tough part is not in understanding the nature of pain, but rather identifying pets that experience it. Veterinary

some conditions are more likely to cause cats pain than others. Bone cancers, for example, can be very painful, while blood-formed cancers such as leukemia or lymphoma have little pain associated with them. Acute pain is associated with disease processes such as pancreatitis, gastrointestinal disease, or feline lower urinary tract disease, as well as surgery and trauma. Chronic pain, on the other hand, can stem from musculoskeletal disease, various forms of cancer, and dental problems.

professionals depend on their clients to notice a problem exists and then to do something about it.

Veterinary experts say recognizing pain in cats can be difficult, and each case varies. Unfortunately, the changes caused by pain are often subtle and appear slowly. We may mistakenly attribute these changes to a natural part of aging. Moreover,

To deal with your older cat's pain, you and your veterinarian must team up to identify the source of the discomfort and solve the problem. Your cat's overall health must be evaluated before an intelligent decision on managing pain is reached. In some cases a surgical procedure or treatment can permanently resolve the problem. In others,

where, for instance, degenerative joint problems or arthritis create chronic distress, medications may be prescribed to control or minimize pain.

You may also want to explore alternative therapies for their potential role in treating chronic pain. Acupuncture, for example, has been shown to increase brain endorphin levels and reduce discomfort. In addition, you can take steps to help your older cat deal with chronic pain—or recover from surgery—at home by doing the following:

• Provide tender loving care

• Minimize physical stress—make litter box, food, and water bowls easily accessible

• Provide easy access to favorite perching areas

• Yield to a hurting cat's natural tendency to hide—provide her with a warm, dimly lit place to sleep comfortably.

HOSPICE CARE FOR CATS

A new, welcome idea for cats and their guardians, hospice care for animals—similar to its counterpart for people—encompasses a mix of practical and emotional help supported by love and compassion. No cat should die alone, in pain, or with strangers. Proponents of hospice care say that the passage through death is sacred and in the end, as wonderful as life. While death and dying can be difficult and emotionally trying, especially for pet guardians who must make a life-or-death decision concerning their animal companion, ultimately it should be viewed as a natural part of life.

Providing hospice care for cats often means simply being there for someone who faces the loss of a beloved pet companion. Don't hesitate to ask others for help as well. People are often relieved to have a specific answer to the statement, "Please let me know if there is anything I can do to help." Here are a few suggestions.

Lend a hand. Daily activities can be overwhelming when someone is faced with caring for a chronically ill or dying pet. Assist with simple tasks, like visiting a friend's home to bring a hot cup of tea, coffee, or lunch. Offer to run errands so the pet guardian can stay home with her cat.

Listen. Sit quietly and let the other person share memories. Offer advice only when asked.

Offer moral support. Volunteer to accompany someone to the veterinarian's office. Even better, offer to drive.

Provide a referral. Offer the phone number of a pet-loss hotline or local support group.

Finally, consider these words from a multi-pet guardian who cares for her chronically ill cat, Cleo. "Cleo has lymphosarcoma and was supposed to be dead a year and a half ago. Thinking this event would happen at any time, I have gone above and beyond in telling her how special she is, how loved, and what a magnificent being she is. I make sure everything is as perfect for her as possible—fresh bedding every day, a good and tasty selection of food, spending extra time with her, and playing music for her—often a Gregorian chant. I know her disease could internalize at any time, so I make sure every minute is quality time for her. I don't dwell on her disease in my mind or with her. I just support her living as best I can."

CHAPTER TWELVE

Endings and New Beginnings

Elder cats give us the gift of time when no one else has any to spare. Like elder humans, they touch us with their fragility, wisdom, sweetness, and comforting warmth. They curl up close by to be near us. They lay on us when we feel sick or depressed. Old cats ask for little in return, perhaps craving only a little more affection as the years pass. By example, they share basic lessons about life and death—that it is okay to slow down and smell the catnip, that we have no choice but to grow old, too. Older cats teach us about inner listening and watching. They spend more hours sleeping. Some believe they begin their journey with small trips to the other side when they sleep. In the end, senior cats teach us patience and courage, how to let go gracefully and, finally, how to age and die with grace.

Facing the inevitable loss of a feline family member is a little like savoring the warmth and waning beauty of an Indian summer—time grows short, yet the season remains lovely and precious. Suddenly we pay attention to smells, sounds, and colors. We appreciate what is all around us because it cannot last. It ends too soon.

Most of us outlive our feline companions. Being blessed with the gift of many years with a beloved cat does not make it easier to let go or say goodbye when the time comes. It is a bittersweet time, not always understood by those who haven't been blessed with the enduring company of a cat.

CHOOSING THE TIME

Euthanasia is the act of causing death without pain. It is a humane option for terminally ill cats or for those with a poor quality of life that cannot be improved through medical intervention. Your veterinarian can provide you with the information you need to make an informed decision and may help you plan for the eventual loss of your pet. Veterinary hospitals often set aside a private room where family members can share a pet's last moments, then stay as long as they wish. Many mobile veterinarians conduct euthanasia in the pet's home, which can be comforting, and comfortable, for all concerned.

Many people choose to hold or touch their pet during the euthanasia process. You may want to sing or speak quietly to your cat, reassuring him of your presence and love, and perhaps letting him know that it is okay to go.

UNDERSTANDING GRIEF

The bond we share with our older cats is deep and wonderful. We have been blessed with many years together, sometimes so long that we do not even remember life without our cat. Ironically, this can also set us up for a deeper level of grief when our beloved pet dies or we choose euthanasia to peacefully end our cat's life. An overwhelming sense of sadness, as well as guilt and loneliness, may pervade our days. The feeling that other people, often those who don't have pets, do not understand what we are going through may make us feel isolated.

DO CATS MOURN?

Those who live with more than one cat, as many cat lovers do, are not alone in their grief. Animal behaviorists agree that cats often mourn when a companion animal dies. They report that cats may become depressed, or stop eating and grooming. The cat may wait in a mutually familiar place day after day, waiting for a companion to return. The grieving can be short, or may last for weeks, depending on the cat and relationship.

In 1996, the American Society for the Prevention of Cruelty to Animals (ASPCA) conducted a Companion Animal Mourning Project, surveying the caregivers of 165 pets. The study found that 65 percent of the cats surveyed showed four or more behavioral changes after losing a pet companion. Seventy percent of cats changed the amount of vocalizations—meowing more or less. Nearly 50 percent ate less and in a few extreme cases, the cats starved.

Author Rita Reynolds recalls the following story: "Years ago, we had two cats, Thomas and Benjamin. They were inseparable. One day we came home to find them 'spread out' on the neighbor's lawn. Both cats looked dead, but as we approached we noticed that only Benjamin was gone, apparently struck by a car. Our neighbor had kindly taken his body out of the road. Thomas was lying right next to Benjamin, his front leg draped over his body in a loving embrace. When we removed Benjamin's body to bury it, Thomas followed, watching over him the whole time. He clearly mourned the loss of Benjamin."

DEALING WITH LOSS

When the time comes, take time to say good-bye. Acknowledge your grief, which may equal that associated with the loss of a person. Allow yourself to grieve and keep the following in mind:

- Realize it is okay to grieve in your own way.

- Understand that guilt is a normal by-product of grief. Responsible caretakers often feel the guiltiest.

- Talk to your veterinarian if you have questions, doubts, or worries concerning your cat's death. Many times a vet can give you a concrete explanation and lessen guilt feelings.

- Be honest with your children. Answer their questions and concerns in an age-appropriate way.

- Honor your cat's life in a way that brings you comfort and peace. You may choose to put together a memory book or photo album, or hold a funeral.

- Join a pet-loss support group, or call a pet-loss hotline.

Learn all you can about a new cat
before introducing her into your home.

STARTING OVER

Bringing a new cat into the family can never replace a beloved older cat. Don't rush into adopting a new pet. Take time to heal, and let some time pass. Wait until you feel ready, or until fate once again scratches at your door. Many cat lovers never look for a new cat; the right cat just seems to find them.

But where can you find a new cat or kitten? Shelters, rescue groups (including purebred rescue), reputable breeders, your veterinarian, a friend, responsible pet stores, and perhaps your own doorstep are all viable options. Keep in mind that animal shelters and rescue organizations throughout the country are full of adult cats who long for good homes. And, as you already know, senior cats have plenty to offer.

While it is easy to fall in love with a cat or kitten that needs a home, be patient and remember you owe it to your potential pet to choose wisely. Learn all you can about a new cat before introducing her into your home. If possible, spend time with her in a private room to see how she interacts with you and your family. Whether finding your cat is the result of research or fate, with proper care and a

lot of love, the end of your search will be the beginning of a beautiful, long relationship once again.

Of course, cats, like people, aren't perfect. Don't discount a kitten or cat you like for minor medical problems. Fleas, ear mites, and worms may make your cat seem a little less desirable, but such problems are easily solved with professional veterinary help. Visit a veterinarian to treat the cat and get her started on a long, healthy life with you.

Spotting a Healthy Cat

Examine your potential cat or kitten before you sign adoption papers. A healthy kitten has the following qualities:

Eyes are bright and clear

Ears are clean and free of discharge

Nose is clear and free of discharge

Mouth has clean teeth and pink gums (unless naturally pigmented)

Coat is smooth, soft, and clean

Anal area is without discoloration or dried waste

Body is lean but not skinny, no potbelly

Pet-Loss Support Hotlines

The following pet-loss hotlines can offer support and a listening ear. Best of all, the people on the other end of the line truly understand what you're going through.

University of California, Davis
(800) 565-1526 or (530) 752-4200

University of Florida, School of Veterinary Medicine
(800) 798-6196

Michigan State University
(517) 432-2696

Chicago Veterinary Medical Association
(630) 603-3994

Virginia-Maryland Regional College of Veterinary Medicine
(540) 231-8038

Ohio State University School of Veterinary Medicine
(614) 292-1823

Tufts University School of Veterinary Medicine
(508) 839-7966

Cornell University Pet-Loss Support Hotline
(607) 253-3932

Iowa State University
(888) 478-7574

Colorado State University Veterinary Teaching Hospital
(970) 491-1242

Resources

ASSOCIATIONS AND ORGANIZATIONS

American Cat Fanciers Association (ACFA)
P.O. Box 203
Point Lookout, MO 65726
(417) 334-5430

American Association of Feline
Practitioners (AAFP)
(professional organization of
veterinarians who share an interest
in providing excellence in care and
treatment of cats; supports
American Board of Veterinary
Practitioners (ABVP) certification
in feline practice)
200 4th Avenue N., Suite 900
Nashville, TN 37219
(615) 259-7788
(800) 204-3514

American Pet Products
Manufacturers Assoc., Inc.
255 Glenville Road
Greenwich, CT 06831
(203) 532-0551

American Veterinary Medical
Association
1931 North Meacham Road,
Suite 100
Schaumburg, IL 60173
(800) 248-AVMA

Animal Protection Institute (API)
P.O. Box 22505
Sacramento, CA 95814
(916) 447-3085

ASPCA Animal Poison Control
Center
1717 South Philo Road, Suite 36
Urbana, IL 61802
(888) 4 ANI-HELP
(888)-426-4435

Association for Pet Loss and
Bereavement, Inc. (APLB)
(worldwide clearinghouse on pet
bereavement)
P.O. Box 106
Brooklyn, NY 11230
(718) 382-0690

Australian Cat Federation, Inc.
(ACF)
P.O. Box 3305
Port Adelaide, SA 5015
Australia
08-8449-5880

Canadian Cat Association (CCA)
289 Rutherford Road South, Unit 18
Brampton, Ontario
Canada L6W 3R9
(905) 459-1481

Cat Fanciers' Association (CFA)
(largest registry of pedigreed cats)
P.O. Box 1005
Manasquan, NJ 08736
(732) 528-9797

Cornell Feline Health Center
(veterinary medical specialty center
devoted to improving the health and
well-being of cats)
College of Veterinary Medicine
Cornell University, Box 13
Ithaca, NY 14853
(607) 253-3414
Consultation: (800) 548-8937,
M,W,F 9 am to 12 pm, and 2-4 pm
(EST)

Reiki Alliance
(to find a Reiki practitioner in your
area)
P.O. Box 41
Cataldo, ID 83810
(208) 783-4848

Royal Society for the Protection of
Cruelty to Animals (RSPCA)
Animal Centre
Weoley Castle
Birmingham B29 5UP
www.rspca.com

ONLINE RESOURCES

American Animal Hospital
Association (AAHA)
www.healthypet.com

American Association of Feline
Practitioners (AAFP)
www.aafponline.org
American Cat Fanciers Association
(ACFA)
www.acfacat.com/

American College of Veterinary
Nutrition (ACVN)
www.ACVN.org

American Pet Products
Manufacturers Assoc., Inc.
(APPMA)
www.appma.org

American Veterinary Medical
Association (AVMA)
www.avma.org

Anaflora
(flower essence therapy for animals
& animal communication)
www.anaflora.com

Animal Protection Institute (API)
www.api4animals.org

Association for Pet Loss and
Bereavement (APLB)
www.aplb.org

Australian Cat Federation, Inc.
(ACF)
www.acf.asn.au/

Canadian Cat Association (CCA)
www.cca-afc.com/

Cat Fanciers' Association
www.cfainc.org

Cat Fancy On-Line Feline Library
www.animalnetwork.com/cats
Cat Massage
www.catmassage.com

Cornell Feline Health Center
http://web.vet.cornell.edu/Public/FH
C/FelineHealth.html

directline.com
(pet policy information in the UK)
www.directline.com

Federation Internationale Feline
(FIFE)
(international cat fanciers society
with members in 40 countries)
www.fifeweb.org/

Feline Advisory Bureau
(promotes the health and welfare of
cats; offers more than 50 information sheets on feline diseases, behavior, and breeding)
www.fabcats.org

Feline CRF Information Center
(Chronic Renal Failure)
www.felinecrf.com

insure.com
(pet policy information)
www.insure.com/personal/pets.html

The International Cat Association,
Inc. (TICA)
www.tica.org/

MyPetTribute.com
(mature pet articles, products, and
services)
www.mypettribute.com

Pet Food Institute (PFI)
www.petfoodinstitute.org

Tellington TTouch
www.TellingtonTouch.com
(Check 'Directory of Practitioners'
to find your nearest TTouch practitioner)

Suggested Reading

BOOKS

Ballner, Maryjean. *Cat Massage: A Whiskers-To-Tail Guide to Your Cat's Ultimate Petting Experience*. St. Martin's Press, 1997.

Callahan, Sharon. *Healing Animals Naturally with Flower Essences and Intuitive Listening*. Sacred Spirit Publishing, 2001.

Commings, Karen. *The Cat Lover's Survival Guide: Helpful Hints for Solving Your Most Pesky Pet Problems*. Barron's, 2001.

Commings, Karen. *Shelter Cats*. Howell Book House, 1998.

Hoffman, Matthew (editor). *PetSpeak: Share Your Pet's Secret Language!* Rodale Inc., 2000.

Mammato, Bobbie, D.V.M, M.P.H. *Pet First Aid*. The American Red Cross and The Humane Society of the United States (HSUS), 1997.

McKay, James E. *Comprehensive Health Care for Cats*. Creative Publishing International, 2001.

Messonnier, Shawn, D.V.M. *Natural Health Bible for Dogs & Cats*. Prima Publishing, 2001.

Moore, Arden. *50 Simple Ways to Pamper Your Cat*. Storey Books, 2000.

Pitcairn, Richard H., D.V.M, and Susan Hubble Pitcairn. *Dr. Pitcairn's Complete Guide to Natural Health for Dogs & Cats*. Rodale Inc., 1995.

Reynolds, Rita. *Blessing the Bridge: What Animals Teach Us About Death, Dying, and Beyond*. NewSage Press, 2001.

Schwartz, Cheryl, D.V.M. *Four Paws Five Directions: A Guide to Chinese Medicine for Cats and Dogs*. Celestial Arts Publishing, 1996.

Siegal, Mordecai (editor). *The Cornell Book of Cats, A Comprehensive and Authoritative Medical Reference for Every Cat and Kitten (2nd edition)*. Villard Books, 1997.

Shojai, Amy. *New Choices in Natural Healing for Dogs and Cats*. Rodale Inc., 1999.

Shojai, Amy. *The Purina Encyclopedia* of Cat Care. Ballantine Books, 1998.

Thornton, Kim Campbell and Calloway, Jane. *Cat Treats*. Main Street Books, Doubleday, 1997.

MAGAZINES AND NEWSLETTERS

Cat Fancy
P.O. Box 6050
Mission Viejo, CA 92690
(949) 855-8822

Cat World
Avalon Court, Star Road,
Partridge Green,
West Sussex
RH13 8RY England
(507) 288-2430 (U.S.)
+44 (0) 1403 711511 (U.K.)

Catnip Newsletter
Tufts University School of Veterinary Medicine
P.O. Box 420234
Palm Coast, FL 32142-0234
(800) 829-8893

Cat Watch Newsletter
Cornell University College of Veterinary Medicine
P.O. Box 420235

Palm Coast, FL 32142-0235
(800) 829-8893

laJoie
(quarterly publication dedicated to promoting appreciation for all beings)
laJoie and Company
P.O. Box 145
Batesville, VA 22924
(540) 456-6204

The Whole Cat Journal
P.O. Box 420235
Palm Coast, FL 32142-0235
(800) 829-8893

VIDEOS

Your Cat Wants a Massage by Maryjean Ballner and Champion (Tape Worm Studios, 1999). For more information, call 1-877-Meow-Meow.

Pet Emergency First Aid: Cats by Apogee Communications. Approved by the American Society for the Prevention of Cruelty to Animals (ASPCA). For more information, call 1-888-380-9966, or visit www.apogeevideo.com/cats/

Index

CPR, 88–93

D

death, of cat, 122, 125
    euthanasia, 125
    grief about, 127
    support hotlines, 130
dental hygiene, *see* teeth
detoxifiers, 40
diabetes mellitus, 108–109
diet, *see* nutrition
digestive system, 30
dislocations/fractures, 86

E

Ear TTouch, 76
ears
    aging of, 28
    examining of, 19
elimination, *see* bowel movements; urinary system
emergencies, *see* first aid
endocrine system, 31
euthanasia, 125
exercise, 44, 46, 49
    toys for, 48
    weight loss and, 51
eyes
    aging of, 28
    examining of, 20
    nutrition and, 37

F

Feline Leukemia Virus (FeLV), 112
fever, *see* temperature, of cat
first aid, 82
    basics of, 84
    for bleeding, 85
    for burns, 86
    for choking, 85–86
    CPR, 88–93
    for fractures/dislocations, 86
    for frostbite, 86–87
    kit for, 88, 95
    kit for homeopathic, 68
    for poisoning, 87–88

    safety tips, 90
    for shock, 85
flea products, poisonous, 88
flower essences, 72, 75
flowers, poisonous, 87
foods, *see* commercial foods; nutrition
fractures/dislocations, 86
frostbite, 86–87
fur
    aging of, 29
    clipping of, around wound, 86
    examining of, 19

G

gases, poisonous, 88
gastrointestinal system, 30
glucosamine sulfate, 40
Gray, Alexander, 15
"green" foods, 40
grief, 127, 130
grooming
    by cats, 29
    of cats, 56–58
gums
    care of, 57
    common problems of, 98–99
    examining of, 19

H

hair balls, 56
health insurance, 113
health problems, common, 96–97. *See also* first aid; illness;
    natural medicine; preventive care
    arthritis, 98
    cognitive dysfunction syndrome (CDS), 102
    constipation, 102
    incontinence, 99
    with litter box, 99
    personality change, 102
    with teeth and gums, 98–99
health, signs of good, 128
Healthy Pets 21 Consortium, 11
hearing loss, 28, 56
heart, *see* cardiovascular system
Heimlich maneuver, 86

Photo Credits

Keith Brofsky: 6, 52, 80

Courtesy of Cape Ann Animal Aid Association, Inc.: 79, 94, 100

Janis Christie / PhotoDisc / PictureQuest: 111

Corbis / PictureQuest: 6, 7, 8, 10, 14, 23, 25, 29, 63, 96, 103, 123

DigitalVision / PictureQuest: 6, 35, 122, 129

Courtesy of Jill Feron: 108

Bill Gallery / Stock, Boston Inc. / PictureQuest: 124

Courtesy of The Glidden Company: 53

Courtesy of Regina Grenier: 50, 59, 65, 77

G.K. & Vikki Hart / PhotoDisc / PictureQuest: 1, 3, 112, 104

Weems S. Hutto: 57, 70, 74

Kent Knudson / PhotoLink / PhotoDisc / PictureQuest: 83

Ryan McVay / PhotoDisc / PictureQuest: 6, 7, 12, 17, 18, 32, 34, 43

Clement Mok / PictureQuest: 26, 55

PhotoDisc: 27, 28, 71, 89, 116, 117, 131

PhotoLink / PhotoDisc / PictureQuest: 114

Tony Ruta / Index Stock Imagery / PictureQuest: 66, 97, 127

Courtesy of Holly Schmidt: 16

Courtesy of Barbara States: 60, 62, 67, 119, 125, 132

Stock South / PictureQuest: 105

Betsy Stowe: 37, 38, 107, 120

Michael W. Thomas / Focus Group / PictureQuest: 45

Sandra L. Toney: 22, 44, 46, 49

About the Author

Susan Easterly is an award-winning columnist and contributing editor to *Cat Fancy*, a popular national magazine dedicated to cat care for the responsible owner. She has contributed to several books on pet care, including *New Choices in Natural Healing for Dogs* and Cats (Rodale, 1999) and *Petspeak, Share Your Pet's Secret Language!* (Rodale, 2000). She is the author of *The Guide to Handraising Kittens* (TFH, 2000), a 2001 Certificate of Excellence winner from the Cat Writers' Association. She has written hundreds of pet articles in magazines and online, ranging from *Modern Maturity* and *KittensUSA* to Pets.com and the Popular Dogs series.

Easterly holds BA degrees in journalism and English (creative writing). She frequently covers animal welfare issues and could not imaging living without a cat or dog. She lives in Newberg, Oregon, with her family and six pets. Her companion animals sometimes find themselves described in Easterly's articles but they aren't alone. The late Cleveland Amory refers to Easterly and her tongue-in-cheek article about giving a cat a pill in his international best seller, *The Cat Who Came for Christmas*.